South Africa's Armaments Industry

Continuity and Change after a Decade of Majority Rule

Dan Henk

University Press of America,® Inc.
Lanham · Boulder · New York · Toronto · Oxford

4501 Forbes Boulevard
Suite 200
Lanham, Maryland 20706
UPA Acquisitions Department (301) 459-3366

PO Box 317
Oxford
OX2 9RU, UK

Printed in the United States of America
British Library Cataloging in Publication Information Available

Library of Congress Control Number: 2006923337
ISBN-13: 978-0-7618-3481-6 (clothbound : alk. paper)
ISBN-10: 0-7618-3481-8 (clothbound : alk. paper)
ISBN-13: 978-0-7618-3482-3 (paperback : alk. paper)
ISBN-10: 0-7618-3482-6 (paperback : alk. paper)

™ The paper used in this publication meets the minimum requirements of American National Standard for Information Sciences—Permanence of Paper for Printed Library Materials, ANSI Z39.48—1984

Contents

List of Illustrations

FIGURES

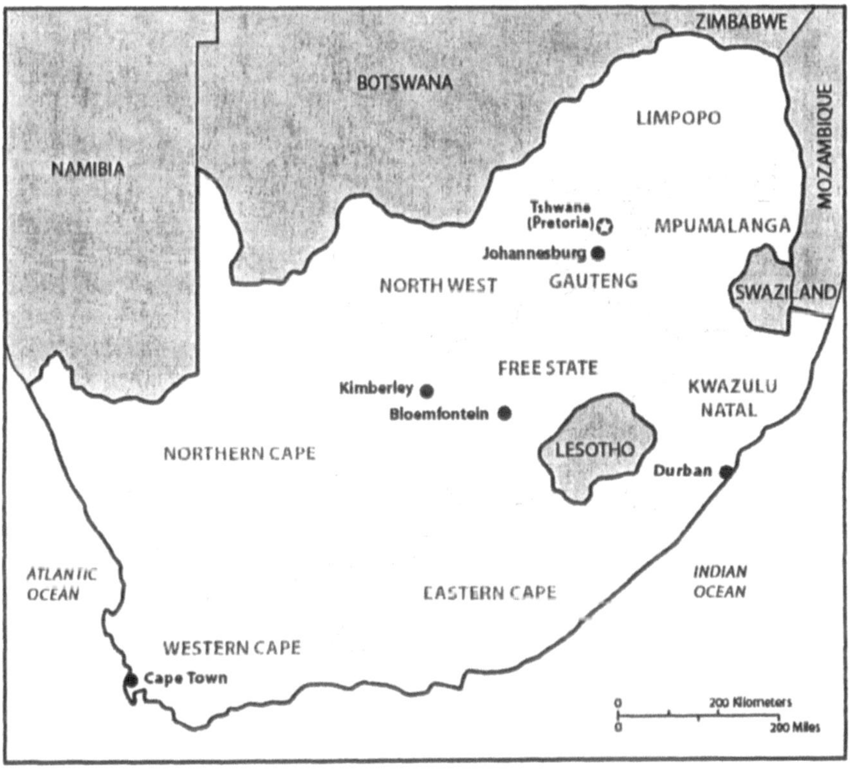

Figure P.1. South Africa in 2005

Preface

Ex Africa semper aliquid novi (out of Africa always something new) wrote Roman scientist Pliny the Elder in classical times, and from his day to ours, Africa has consistently astonished outsiders. The wonders of its natural environment and the richness of its resources have long been irresistible lures, even as the fierceness of its tropical diseases cut short the lives of visitors and options of inhabitants. Africa's huge land area, along with its extraordinary environmental and cultural diversity, almost inevitably reduces conceptions of the continent to stereotype and generalization—the magnitude and complexity of African detail defy easy categorization.

Pliny's aphorism has proven unfortunately true of African tragedy in recent times as well. Its people have endured the worst that man or nature can devise, horror piled upon unbelievable horror. Depictions of Africa in Western media range from emaciated victims of searing drought to axe-wielding, drug-crazed psychopaths intent on excising life and limb of helpless children. Diseases stalk the continent's media landscape, whether the quick horror of hemorrhagic fevers, the cholera of refugee camps, or the gruesome, slow death of acquired immunodeficiency syndrome, AIDS. Economic distinctions also are painful: visitors to African cities see wealthy playboys leering down from billboard advertisements on urban youth traumatized by grinding, hopeless poverty. Suffering seems to be one of the few African constants. Yet arguably, no continent has surpassed this one in triumphs of the human spirit. Set in sharper relief against a background of calamity and misery, the qualities of hope, compassion, reconciliation, and cooperation shine very brightly. If Africa's misfortunes can stun the rest of the world, its triumphs can equally inspire the world. In an era saturated with stories of exploitation and loss, Africa affords uplifting instances of transcendence. This is one such story.

The Industrial Revolution arrived episodically and late on the African continent, but one locale where it was firmly entrenched by the middle of the twentieth century was South Africa. Here, as elsewhere, industrialization produced deep cleavages of social class. However, class distinctions in South Africa were intermixed with profoundly divisive racial categorizations. These ultimately contributed to virulent conflict, as a small ruling elite of European ancestry exercised its exclusive advantages to build a powerful security establishment and sophisticated arms industry to protect its privileged position, employing modern arms with devastating effect. Yet ironically, the industry that once empowered an unrepresentative elite now provides substantial economic benefit to a multi-racial, democratic country, furnishing South Africa's government with instruments of national power unrivalled elsewhere in sub-Saharan Africa. This book tells the story of the country's arms industry, emphasizing particularly the remarkable process by which it transformed itself in the minds of the country's citizens from the coercive instrument of minority rule into a valuable asset for the nation as a whole.

There is undeniable irony in this story, argued here as a triumph of the human spirit, involving an industry producing items designed to destroy and kill. The sensibilities of some readers may not admit of the possibility that arms can be a valued servant as well as a cruel master, or that the same materiel may serve equally to kill, protect, or rescue. Here, the classical thinkers may usefully inform the thinking of the present age. Were he resurrected in our day, Aristotle undoubtedly would observe that humankind is no closer to eliminating violent aggressors or the ravages of nature than were his contemporaries; he would urge us to the *phronesis*, the mature wisdom that combines science, morals, and experience to provide the right answers to the security questions of the time. Thucydides likely would point us to the wars of his Greek contemporaries and warn us that security cannot depend on political systems and astute alliances. Though perhaps controversial in our day, the observation of the fourth century Roman Vegetius Flavius Maximus also has a certain undeniable resonance: *"nemo provocare . . . quem intelliget superiorem esse pugnaturem"* (no one provokes he who is known to be superior in battle). More to the point, in the contemporary notion of the social contract, one of the most fundamental responsibilities of the state is to secure the citizenry from a range of threats. Accountability for that security falls in large measure on a nation's armed forces, which depend in turn on technology for capability and credibility.

When the author began the research for this book, South Africa's political transition was not his prime concern. Rather, the initial interest was the country's military technology itself, and two underlying questions were: (1) What armaments are produced in South Africa? and (2) How sophisticated are they? That initial focus, however, was quickly replaced by fascination with a broader issue: the remarkable story of the rehabilitation of the nation's arms industry in the 1990s to serve the interests of the new South Africa. Political changes in that period included a fundamental South African rethinking of "security." And in the new security construct, the country had discovered a useful role for its arms industry—an important but potentially problematic pillar of the *ancien régime.* By the early years of the twenty-first century, South Africa had clearly begun to weave its arms industry into the fabric of the country's future. How (and why) it accomplished this feat is a fascinating story with unambiguous relevance for other societies.

Beyond its possible applications, there is at least one other compelling motivation for reconstructing the story at this time. A surprisingly large number of its important participants are still available for interview—many remain active in South Africa's government or industry—and their memories comprise an invaluable trove of perishable detail. When the current generation passes from the South African scene, it will be difficult to capture some of the most salient elements of the story, particularly those transactions in human relationships that define the most directly applicable lessons. In fact, the granularity of that detail is essential to understanding the full dimensions of the country's achievement. It is the author's fervent hope that he has captured some of it for posterity and that these efforts will encourage other scholars to examine the story more closely.

This book leaves many unanswered questions. Readers interested in technological detail will find only a modest treatment of the specific products of the defense industry and may be left wondering just what astonishing new weapons are about to come off South Africa's assembly lines. Some readers will be disappointed to find no exhaustive analysis of South Africa's military technology and weapons engineering. Still others interested in industrial management may find themselves pining for more discussion of the leadership principles and philosophies of individual South African industry leaders. Others will seek in vain for detailed economic analyses of corporate performance. Some will look for details in the scandals that have plagued South Africa's arms acquisitions. These all are legitimate interests, of course, but their elucidation must be sought elsewhere. The focus of this book is the general social and political process that mediated a significant transition of an armaments industry in a developing country. The relevant questions are:

(1) What changed?

(2) Why?

(3) What is the likely vector of future change?

(4) What can be learned?

Many people contributed to this book, providing the information and insights reflected on its pages. It is impossible to do justice to them all, but several individuals deserve to be singled out for their particularly valuable contributions. Foremost among them is Helmoed-Römer Heitman, South African correspondent for *Jane's Defence Weekly*, renowned chronicler of regional military affairs, and one of the most astute military analysts anywhere. Len Le Roux lived through South Africa's political transition as a senior military insider and key actor in his own right. A deeply reflective analyst, Le Roux could offer insights available from few others. Paul Gerber of Grintek and Tony Savides of BAE Land Systems OMC, South African arms-industry insiders and retired senior military leaders, drew on the extensive experience in their military and industrial careers and were extraordinarily helpful and generous of their time and perspectives. Alan Holloway, a long-time arms-industry executive and now an industry captain, displayed patience, wit, and generosity in addition to marvelous historical perspective and profound analytical insight. Gavin Cawthra, scholar and student of security issues at Witwatersrand University, participant in the anti*apartheid* struggle and actor in the consultations that produced the new South Africa, generously shared his experiences and deep knowledge of the political context. Johan van Rensburg, commercial analyst at the US Consulate in Johannesburg, offered highly useful perspectives on current industry trends and US-South African relations. Duncan Lang, who works for the US Secretary of Defense, encouraged the research and also contributed valuable detail on US-South African arms-industry connections.

Other contributors were Daniel M. Armstrong of the Air University Press and Jack Durham of J. Durham LLC, who drew the maps that appear in this publication. Several individuals read drafts of the book, suggesting improvements and calling attention to oversights, errors, redundancies, and faulty analysis. These included Helmoed Heitman, Alan Holloway, Len Le Roux, Paul Gerber, Tony Savides, Martin Rupiya, Abel Esterhuyse, and Johan van Rensburg in South Africa. Readers in the United States included Ed Burkardt, John Geis, Grant Hammond, Jeff Hood, Jim Winkates of the US Air Force's Air War College, and Duncan Lang in the Office of the Secretary of Defense. Any merits this book may have are due in large part to the generosity of these contributors. Errors of fact and defective analysis are, alas, the sole responsibility of the author.

Introduction

In November 1998, South Africa announced its intention to purchase a variety of sophisticated new weapons from a select list of foreign suppliers.[1] The following year, it signed contracts for jet aircraft from the United Kingdom and Sweden, patrol corvettes and submarines from Germany, and utility helicopters from Italy. These acquisitions had an initial value of about four billion dollars.[2] They were controversial in a nation struggling to meet many human needs and facing no immediate external security threat. With subsequent allegations of graft and kickbacks, public dissatisfaction was widespread and occasionally harsh.[3]

Despite the controversy among South Africans, the weapons purchases were not the arbitrary choice of a corrupt autocrat, nor were they indicative of an out-of-control military establishment. Rather, they were one step in a remarkable process of democratic decision-making following a rational assessment of long-term defense needs. They reflected deliberate strategic choices and prioritization of national interests. They also illustrated one country's struggle both to define its security ends and its choice of the ways and means to achieve them. And they confirmed South Africa's determination to retain state-of-the-art "strategic capabilities."

While South Africa has a large and sophisticated armaments industry, its decision to spend billions of dollars on foreign arms might at first appear to be an explicit rejection of homemade products. However, that conclusion would be mistaken. Even though much of South Africa's new weaponry would come from abroad, its future strategic capabilities would not depend on foreign suppliers; for ironically, its choices also indicated a South African intention to maintain a sophisticated, indigenous armaments industry. Much of the technology in the new weapons actually would be designed and produced locally, and through the partnerships that ensued, the arms acquisitions would enable South Africa to sell its military technology around the world as components of sophisticated European weapons.[4] Even in selling their arms, the foreign producers ultimately would invest more money in South Africa than they received in compensation for their products.[5] The country's entire economy would profit from the purchases, and no sector more so than the defense industry.

The very existence of an arms industry in South Africa is an interesting phenomenon. It is relatively young – a product of the latter 20th Century – and almost unprecedented in rapidity of development. It was virtually nonexistent in 1960. Yet twenty-five years later it employed 150,000 people, was producing most of the materiel needed by South Africa to fight its regional wars, and included highly sophisticated state-of-the-art applications, ranging from self-propelled artillery to electronic warfare to nuclear weapons. Arms also had become a critical source of foreign exchange. But all this was achieved at a very high cost.

South Africa's history in the latter half of the 20th Century is complex and fraught with circumstances of wrenching social change. For the three decades prior to 1990, the country's decisions about "security" were the prerogative of a small group of relatively unaccountable senior officials. Until 1989, this meant South Africa's exclusion of the majority of its population from opportunity and access, a struggle against liberation movements in the region, and a participation in the front lines of the Cold War against the Soviet Union and its proxies. The quest for this kind of security made South Africa one of the handful of states in the developing world willing to bear the economic burden of developing not only a massive conventional arms industry but also nuclear weapons and a space program (both of the latter dismantled in the early 1990s). Still, South Africa's recent history bequeathed it an armaments industry like no other in Africa.

The country's political past is a feature of recent memory. An ethnic minority of European descent ruled the country from the time it took shape in its present form in 1910 to the establishment of majority rule in 1994. The National Party, in power from 1948 to 1994, formalized a system of coercive, institutionalized racism under the rubric of ethnic "separate development" (*apartheid*) that denied non-whites basic rights and opportunities. This policy ultimately drove many South Africans into anti-regime activity and armed struggle.[6] Those policies were given a very public face in the massacre of seventy black protestors by South African security forces at Sharpeville in 1960.

South Africa's racial policies and repression of domestic dissent increasingly alienated the rest of the world after the mid 1950s. By the early 1960s, the Communist Bloc had begun to support indigenous liberation movements in southern Africa, while white-ruled South Africa opposed these movements and endeavored to maintain an alignment with the anti-communist West. From the 1960s onwards, the country fought against its own internal dissidents and figured prominently in regional Cold War struggles, backing other white minority regimes in southern Africa until, one by one, each succumbed to majority rule. South Africa then sponsored insurgencies against the Marxist governments of neighboring black-ruled states.[7] By the late 1970s it was engaged in protracted regional conflict, ranging sporadically along a scale from low to mid-intensity, and involving the country's ground, air and special forces. The country was fighting both its own dissidents and regional Soviet clients that included conventional forces and insurgent groups based in neighboring states.[8]

South Africa had withdrawn under pressure from the British Commonwealth in 1961. In 1963, the UN imposed a "non-mandatory" embargo on arms sales to South Africa, and in late 1977 a "mandatory" one.[9] Neither UN embargo cut the country off completely from the international arms trade, but international sanctions after 1963 increasingly reduced South Africa's options and significantly augmented the cost of acquiring military materiel.[10] The region saw something of an arms race in the 1970s and 1980s, as South Africa strove in the face of international sanctions to match or exceed the military capabilities of its Soviet-backed foes. The growth of the defense industry, and its present features, are very much a product of this arms race.

Figure I.1. Southern Africa in the Early 1960s

Until the 1970s, South Africa maintained relatively overt military relationships with countries such as France, and even in the wake of the mandatory embargo in 1977, successfully pursued covert bilateral security and technology exchange programs with other "pariah" states such as Israel, Chile and Taiwan.[11] By the 1970s, South Africa also was heavily engaged in tortuous and elaborate covert operations to obtain industrial intelligence and materiel from the developed western countries. This required concerted efforts to evade the UN embargoes through front companies, smuggling and intellectual property theft.[12] Still, the country's most important arms acquisition program involved an escalating series of initiatives to achieve self-sufficiency in arms production, an endeavor stretching from the 1960s into the 1980s. South Africa continued to depend on its private sector for arms, but by the late 1960s also had created a unique public sector production conglomerate specifically charged with finding or producing the increasingly sophisticated materiel required by its armed forces. This public sector entity – the Armaments Corporation of South Africa (Armscor) – became the most important part of the South African armaments industry in the *apartheid* years.[13]

By the mid 1980s, South Africa had come very close to achieving self-sufficiency in military materiel, producing an incredibly wide range of arms, and displaying an astonishing ability to rapidly translate military requirements into fielded weapons systems. The sophistication of its weaponry gave the country a significant local advantage in its regional wars. By this point, South Africa also was able despite international sanctions to market increasing amounts of its military materiel abroad. However, the country embarked on a dramatic process of political reform in 1989 that seemingly obviated its need for the sophisticated materiel its arms industry had produced over the two preceding decades.

With the advent of majority rule in 1994, South Africa rejected some of the most unfortunate social baggage of its past, but it also faced many difficult choices. One had to do with the nature of the armed forces it would maintain. Still another was the disposition of the highly sophisticated arms industry, an important pillar of a thoroughly discredited former government. Given the diversity of political voices and agendas in South Africa at the time, little could be taken for granted. In addition to its association with the *apartheid* regime in the minds of the country's citizens, the arms industry faced other daunting challenges in the years immediately after the transition to majority rule. It struggled to achieve financial viability and to overcome an unwelcome international reputation for profligate sales to unsavory clients. Yet after a decade of majority rule, the industry not only had survived the political transition, it had established a respected position as a key government partner. It was smaller than it had been in 1989 but was selling an impressive range of products on the world market and was playing an important and growing role in South Africa's economic development and foreign trade.

By the early years of the 21st Century, South Africa had become a global actor in the international arms market. It was carving out significant high-technology niches where its products competed effectively.[14] Nor had the South Africans ignored the security technology of 21st Century humanitarian crises. Its armored vehicles and mine-detection equipment were widely used by the United Nations Organization and other clients. If the eastern United States were subjected to another terrorist strike, state police would respond in sophisticated South African-made emergency vehicles.[15] In fact, there are few regions of the world where the South Africans have not marketed their security-related products.

This book describes how (and why) South Africa acquired an arms industry between 1960 and 1990, then recounts how the industry adapted to survive the political trauma of the 1990s.[16] It ascribes that feat to a remarkable process of consultation amongst key stakeholders in the government, industry, academe and civil society, beginning in the early 1990s, that redefined the role and capacities of the industry. The book calls attention to the emerging partnership between civil authorities, military consumers and industry leaders that now assures indigenous industries will provide "strategically essential technologies" to the country over the long term. It also notes the considerable economic benefits that the arms industry and arms purchases and arms exports are beginning to bring to the nation.

South Africa's technical achievements and its entrepreneurial successes are impressive, but the most significant contribution of this work is its description of a process of change in which members of a society overcame significant historical and ideological obstacles to attain noteworthy agreement about key issues relating to their security. That agreement linked an extraordinary variety of actors and is by no means complete, nor has it been easily achieved, requiring the continuing accommodation of highly diverse interests. A series of scandals also has continued to tarnish both industry and government. Nonetheless, this is the story of a society that overcame the resentments of the past in a deliberate series of collective choices to shape a desirable future. The South Africans demonstrated how an institution created to uphold sectarian interests could be rehabilitated to serve the interests of the larger society. So if this African country with a legacy of conflict between ideologies, classes and races can through broad agreement forge an instrument of a discredited past into a vital asset of a new democracy, surely other societies can follow its lead.

The remainder of the book is organized into five chapters. Chapter 1 examines the arms industry's growth in South Africa from its modest beginnings immediately after World War II to its large size and remarkable capabilities by the mid 1980s, relating that evolution to the South African perceptions of threat and the concurrent impact of United Nations sanctions. This chapter briefly examines the sources of South African arms and arms technology during the *apartheid* era, and notes the development of Armscor as the controlling actor in arms development and acquisition.

Chapter 2 describes the processes of change that mediated the industry's transition in the 1990s from the *apartheid* state to a new, democratic South Africa. This chapter briefly examines the political context of the period and notes the roles of several key actors. It describes the evolution of the concept of "security" in the transition to majority rule, describes the policy instruments that documented and mandated the changes and briefly describes the complex relationship between the government and the arms industry in the new South Africa, in which the government plays the dual role of overseer and partner. This chapter also describes the way in which South Africa's post-*apartheid* leadership has endeavored to avoid scandal in its arms exports.

Chapter 3 illustrates the emerging industry/government partnership by examining two overlapping issues that serve as case studies in the close relations between the South African government and the arms industry – the countertrade protocols and the "strategic packages" of foreign arms purchases announced by the South African government in 1998. This chapter briefly outlines South Africa's engagement with foreign partners in materiel acquisition and arms exports. It also discusses the particularly fraught arms-related connections to the United States between 1994 and 2005.

Chapter 4 offers a summary of the contemporary South African arms industry, noting both the extensive government role in the industry as a whole and the continuing distinction between public and private sector components. This chapter offers a list of "benchmark" products that illustrate the accomplishments of the South African arms industry since the mid 1970s, and analyzes South Africa's security-related technology, identifying its strengths, limitations and the country's success in pursuing niche opportunities.

Chapter 5 concludes the book by reflecting on continuities, ongoing change and the future of the industry. In a list of "lessons learned," it offers an assessment of implications for South Africa, its neighbors and its partners. Appendices at the end of the text offer more detail on specific corporate actors in the country's arms industry.

Notes

1. The announcement was made at a South African arms exhibition (DEXSA-98). "South African gets ready to announce fighter trainer winner," Reed Business Information Ltd., 14 October 1998; "$5.2bn military deal defended against backdrop of regional war," *Financial Times* Information, Southscan, 1 December 1998.

2. David Botha, *Offsetting the Costs of SA's Strategic Defence Packages* (Pretoria: Institute for Security Studies, ISS Occasional Paper 75, 2003), 5. The future costs of the program were estimated the time to be up to an additional US$5 billion. Cabinet Decision on Strategic Defence Procurement, 15 September 2000, http://www.gov.za/papers/procurement/cabinet.htm (5 June 2004). US dollar costs are reflected in the 1999 Rand/Dollar exchange rates.

3. David Botha, "The Arms Deal Controversy," *African Security Review*, 12, no. 3 (2003): 99-103.

4. Peter Batchelor and Paul Dunne, "Industrial Participation, Investment and Growth: The Case of South Africa's Defence Related Industry" (paper prepared for the Centre for Conflict Resolution, Cape Town and Middlesex University Business School, United Kingdom, February 2000), 2.

5. Bothá, *Offsetting the Costs*, 9.

6. See, *inter alia*, Steven Mufson, *Fighting Years: Black Resistance and the Struggle for a New South Africa* (Boston: Beacon Press, 1990); Howard Barrell, *MK: The ANC's Armed Struggle* (London: Penguin, 1990); and Robert Harvey, *The Fall of Apartheid: The Inside Story from Smuts to Mbeki* (New York: Palgrave, 2001).

7. See, for instance, Robert M. Price, "Pretoria's Southern African Strategy," in *Exporting Apartheid: Foreign Policies in Southern Africa 1978-1988*, ed. S. Chan (New York: St. Martin's Press, 1990), 145-69; J. Hanlon, *Apartheid's Second Front: South Africa's War Against its Neighbours* (London: Penguin, 1986); William Minter, *Apartheid's Contras: An inquiry into the roots of war in Angola and Mozambique* (Johannesburg: Witwatersrand University Press, 1994); and Willem van der Waals, *Portugal's War in Angola, 1961-1974* (Rivonia, South Africa: Ashanti Publishing, 1993), 208-12.

8. For overviews of these events from a military perspective, see Hilton Hamann, *Days of the Generals*, (Cape Town: Zebra Press, 2001); for order-of-battle detail, see Helmoed-Römer Heitman, *The South African War Machine* (Novato, CA: Presidio Press, 1985).

9. These were UN Security Council Resolutions 181 and 418 respectively. For a useful overview, see James P. McWilliams, *Armscor: South Africa's Arms Merchant* (London: Brasseys, 1989), 133-7.

10. For detail, see particularly Signe Landgren, *Embargo Disimplemented: South Africa's Military Industry*, (New York: Oxford University Press, 1989), 191-247. For a discussion of the South African reaction to the events of the 1960s, see S. Ratcliffe, "The shifting dominance: the foundations of the armaments industry during the 1960s," *Africa Perspective*, 25 (1984): 24-39.

11. Ibid. See also A. Leysens, "South Africa's military strategic link with Latin America: past developments and future prospects," *International Affairs Bulletin* 15, no. 3 (1991): 23-47.

12. Gavin Cawthra, "South Africa," in *Arms Procurement Decision Making, Volume II: Chile, Greece, Malaysia, Poland, South Africa and Taiwan*, ed. Ravinder Pal Singh (Oxford: Oxford University Press, 2000), 145.

13. In the late 1960s, then Minister of Defence P.W. Botha toured arms industry facilities in Europe, particularly in Portugal and France, in a deliberate effort to define a more sophisticated South African armaments industry. Botha translated his observations and contacts directly into the growing South African defense industry, an agenda he continued to pursue throughout his long tenure as Minister of Defence (1966-1978) and after assuming the office of Prime Minister in September 1978. Landgren, *Embargo Disimplemented*, 45; Peter Batchelor and Susan Willett, *Disarmament and Defence: Industrial Adjustment in South Africa* (Oxford University Press, 1998), 30. For additional contextual detail, see also P. Frankel, *Pretoria's Praetorians* (Cambridge: Cambridge University Press, 1984), 82.

14. These included avionics, electronic warfare and electro-optical applications now installed in the most sophisticated European combat aircraft. The country had achieved export success in other niche markets. In April 2004, the US Army tested a prototype 105mm gun made by the South African firm Denel on its new *Stryker* combat vehicles. By that point the South African parastatal Denel had developed a relationship with the US-based General Dynamics Land Systems (GDLS) to jointly explore new artillery and ammunition technology. In early 2005, the US Army signed a contract with General

Dynamics Land Systems – Canada (GDLS-C) to purchase 148 mine-protected vehicles manufactured by the South African firm BAE Land Systems OMC, along with spare parts, training and support. At the time, this appeared to be the first of a continuing series of US Army purchases of South African equipment for countering or detecting mines. Author's interview with Duncan Lang, contractor to Office of the US Secretary of Defense, 12 March 2004. See also http://www.baelandsystemsomc, 26 March 2005.

15. These are the RG-12 environmentally protected emergency vehicles built by the South African firm BAE Land Systems OMC and marketed in the United States by General Dynamics Land Systems-Canada. Author's interview with Brigadier Tony Savides (South African Army, retired), at the time Director of International Marketing, Alvis-OMC (now BAE Land Systems OMC), 18 June 2003. See also http://www.baeland systemsomc, 26 March 2005.

16. This book will continue to refer to a South Africa arms industry, but the South African government does not view the country's arms producers as a distinct industrial sector, arguing with justification that arms producers are imbedded components of a larger economy and that the same industries sell products to non-military consumers. In describing that portion of the economy that produces armaments, the South African government prefers the phrase "defence-related industries." *White Paper on the South African Defence-Related Industries*, December 1999, 2, http://www.gov.za/whitepaper/1999/defence.htm, (12 June 2004).

CHAPTER 1

Why a South African Arms Industry?

An Early Incentive for Indigenous Arms Production

The roots of the South African arms industry are surprisingly shallow, but the prerequisites for such an industry were present in the country by the mid 20th Century. South Africa had a sophisticated, diversified economy with a powerful industrial sector and a large (and growing) community of engineers and scientists.[1] Its excellent universities were producing a steady stream of graduates in engineering and the hard sciences.[2] Until World War II, few of these resources had been channeled into production of armament, and South Africa looked to the United Kingdom for the materiel to equip its security forces. This may have been an inevitable consequence of the country's colonial history, but it was a situation not destined to last.[3]

South Africa had emerged in 1910 as a self-governing union within the British Empire, a legacy of British colonial expansion in the 19th Century. The Union's political and economic elites in the early 20th Century were sympathetic to metropolitan interests, but the circumstances of the country's origin had left profound divisions within the dominant white population. The advent of a National Party government in 1948 was a catalyst for change.[4]

Though the party had participated in earlier coalition governments, it increasingly had come to represent the more rural, conservative Afrikaner interests in the population – a constituency that nurtured lingering historical resentments and considered itself significantly disenfranchised by previous governments. This core constituency was neither sympathetic to the interests of the United Kingdom nor committed to the Commonwealth. The party leadership in 1948 was not inclined to support the UK's Eurocentric postwar defense policies and at first did not appear particularly interested in a significant defense force.[5] However, the new leaders sensed from the outset that their domestic policies would alienate South Africa from traditional partners and sources of armament, and they began to explore alternatives.

A model for defense production had emerged in World War II, when local industries had been harnessed to Allied war production in a well-organized partnership with the government. Although that relationship lapsed after the war, the government now rehabilitated it. As the 1960s progressed, the South Africans refined the structures and relationships of defense production, perfecting a system that drew materiel from three separate sources: from a growing community of parastatal enterprises, from arms producers in the private sector and from overseas defense industries. The early steps were halting and uncoordinated; but after a quarter of a century of steady progress, the country had created a very sophisticated indigenous arms industry.[6]

Although South Africa left the Commonwealth in 1961, it continued to enjoy close strategic relationships with the United Kingdom and United States until the mid 1970s. South Africa also derived a perception of security from the presence of the white-ruled colonies on its borders. That situation changed after the mid 1960s, as each of the neighbors achieved majority rule. Then, in 1977, the United Nations imposed its "mandatory" embargo on arms sales. The mid 1970s thus represented a watershed period in South Africa's relations with the outside world, reinforcing its sense of strategic isolation, dramatically increasing its fears of geopolitical vulnerability, and strengthening its resolve to achieve self-sufficiency in armament. The single most important contributing event was the South African experience in the Angolan civil war in 1975.

The Impact of the Angola Incursion (1975-1976)

Despite the country's efforts to procure and produce military materiel in the early 1960s, much of its equipment remained comparatively obsolescent into the 1970s. However, the adequacy of that armament was called into serious question in 1975 when South Africa intervened in Angola, a colony now on the brink of independence from Portugal. By this point the Angolan civil war had become thoroughly embroiled in the proxy conflicts of the Cold War.

The Portuguese had been fighting a bitter colonial war in Angola since the early 1960s, ultimately finding themselves opposed by three separate indigenous nationalist groups, each with its own insurgent army and external patrons. Frustration over its seemingly interminable African colonial wars contributed to a military coup in Portugal in 1974 that overthrew the sclerotic Salazar dictatorship, and resulted in an almost immediate decision by the country's new rulers to rid themselves of their African colonies. Portugal subsequently hosted a conference attended by leaders of the competing indigenous Angolan groups and brokered an agreement amongst them in January 1975.[7] However, by March 1975 the agreement had completely broken down in the midst of escalating conflict. In November 1975, Portugal quietly withdrew its officials from Angola, ending four centuries of colonial presence.

With Portugal's imminent departure from Angola, South Africa and the United States threw their support to two of the insurgent groups in a very loose cooperative effort.[8] The Soviet Bloc backed the third faction.[9] By late 1975, the Soviets were conducting a massive airlift to supply their faction with sophisticated military hardware and Cuban ground and air forces. For its part, the United States provided limited military equipment, funding, and advice to its faction. The South Africans deployed a small ground force into Angola, attacking northwards to the outskirts of Luanda in a fast moving operation.[10]

Figure 1.1. South African Intervention in Angola, 1975 – 1976

Soon after the South Africans intervened in Angola in 1975, they confronted well-equipped Cuban expeditionary forces and received an unflattering test of their ground force equipment in mid-intensity warfare. Though they fought well, they found themselves particularly disadvantaged by the Cubans' Soviet-made artillery with its superior ranges, rates of fire and projectile lethality.[11] The intervention in Angola also exposed serious vulnerabilities in military materiel, logistics and organization for combat.

In early 1976, the United States halted its support to the alliance.[12] The South Africans soldiered on until March 1976, but found their position increasingly untenable and ultimately withdrew. However, they left Angola with a profound sense of betrayal by their erstwhile American allies and a growing perception of strategic isolation. The Angola intervention coincided with an overall erosion of support by former strategic partners that reinforced a South African sense of abandonment. In 1975 the United Kingdom abrogated a long-standing naval relationship involving South Africa's Simons Town naval base. An election in the United States in 1976 brought in a new Administration – that of Jimmy Carter – much more critical of South African policies. Then, in 1978, reacting to South African refusal to sign the Non-Proliferation treaty, the United States halted its cooperation with South Africa on peaceful nuclear issues.

South Africa's intervention in Angola in 1975 marked the beginning of almost a decade and a half of war in Angola between the South Africans and the Marxist regime in Luanda, the latter reinforced with Cuban and other East Bloc allies. By 1980 South Africa had established a more or less permanent military presence in southern Angola, and by the mid 1980s it had conducted a recurring series of strikes and military operations against anti-South African insurgents based in that country.[13] It also intervened repeatedly to assist Jonas Savimbi's National Union for the Total Independence of Angola (UNITA), an indigenous Angolan group waging its own insurgency against the Angolan government. This conflict at times reached mid-intensity proportions, featuring intense artillery duels, air-to-air and air-to-ground combat and multi-brigade operations by mechanized and armored forces.[14] South Africa's support to the Angolan rebels lasted until the advent of a comprehensive regional peace agreement in 1988.[15]

South Africa reacted to its Angolan experience in a number of significant ways. One was a reappraisal of policy and a new security strategy.[16] Since the early 1970s, the South African leadership had been consolidating security decision-making under a small group of senior political insiders.[17] This process escalated as the decade progressed. The politically powerful, relatively unaccountable inner circle came to be widely characterized as the government of "securocrats," a term applied particularly to the Administration of P.W. Botha, South Africa's chief of state from 1978 to 1989. The prerogatives of the "securocrats" were strengthened by the Angolan experience.[18]

After 1975 the country's leaders increasingly were convinced of their abandonment by the West in the face of a "total onslaught" by the Communist world, spearheaded by Moscow's subversive agents and supported by proxies in newly independent African states. This conclusion was reinforced in 1976, when the country's disenfranchised black majority rose up in a protracted period of violent unrest.[19] The South African government responded with a "Total National Strategy" that relied heavily on coercive state agencies to maintain security at home and intimidate enemies abroad.[20] Local scholars described the new strategy as "[operating] at political, economic and ideological levels, [affecting] every area of society, and [impinging] on the lives of all South Africans."[21]

The "Total National Strategy" included an assumption that South Africa would receive minimal assistance from any external source, an assumption reinforced in 1977 by the "mandatory" UN sanctions against arms sales to the country.[22] Its new strategy envisioned a significant capacity to attack enemies by air and by special operations forces well beyond the national borders, including the destabilization of hostile neighbors and sponsorship of insurgencies, intentions that required sophisticated ordnance and intelligence.[23] In the 1970s the South Africans also reenergized programs to acquire chemical, biological and nuclear weapons as an ultimate strategic deterrent to the Soviets and their regional proxies.[24] At the same time, they undertook a concerted program to improve their military organization for combat and accelerated their ongoing efforts to field new generations of indigenously designed conventional weapons providing greater firepower, armored protection and mobility.[25]

The arms industry evolved in the 1970s and 1980s to support the new strategy and meet the escalating needs of the South African military, particularly for conflict in Angola. It proved increasingly adept at rapid research, production and fielding of new technology. The technology itself came from a variety of sources. In the early *apartheid* years, some was produced under license from European arms industries, and later modified to suit unique local requirements. Some was inspired by materiel captured from regional foes: the South Africans did not hesitate to "reverse-engineer" Soviet weaponry encountered in Angola. Some new materiel came from the experience of neighbors such as Rhodesia.[26] Still more was covertly transferred from foreign sources in violation of the UN embargos, a feature particularly of relations between Israel and South Africa between the mid 1970s and mid 1980s.[27]

Whatever the origin of the materiel, the South Africans studied it carefully and modified it for their own use. They were able to test small numbers of prototypes in combat and made increasing use of sophisticated local test facilities. They learned to compress research and testing cycles in order to field weapons on short notice. However, their isolation from the rest of the world imposed challenges that they never were able to fully overcome: for instance, while they could produce many components, they simply could not manufacture the highly sophisticated aircraft engines that could give their air force technological parity with advanced Soviet-designs like the MiG-23 in the hands of Angolan, East German and Cuban pilots.[28]

South Africa's incentive to acquire superior military technology was driven by the sophistication of Soviet arms in the region, but also was anchored in the fears of the white population, especially the politically dominant Afrikaners with their strong collective memory of conflict against numerically superior enemies, their perceptions of isolation in a hostile world and their aversion to interference by "outsiders." South Africa's small white population – and limited number of specialists like jet pilots – made the South Africa's political leaders very sensitive to casualties, limiting their military options.[29]

Despite their effective use of non-whites in various military roles, South Africa's leaders were unwilling to fully mobilize the non-white majority to fight their country's regional wars. The *apartheid* era military never exceeded 150,000 personnel, most of which were part-time, white citizen-soldiers.[30] The state could never hope to match the manpower available to its opponents – a fact reinforcing the white community's fear of the *swart gevaar* ("black peril"), an obsessive dread for the loss of life, property, status and ethnic or cultural identity in a sea of black peoples.[31] The country depended heavily on the impressive tactical competence of its Defence Force, but this alone could not guarantee a perpetual advantage. Continued military success with limited casualties required a profound technological edge.

Like the military technology it produced, South Africa's arms industry derived some of its unique features from the cultural peculiarities of its employees. White South Africans took considerable pride in their self-reliance, adaptability, technical skills and military prowess.[32] The industry reflected some of the informal practicality of their society itself, a trait captured in the *Afrikaans* proverb: *'n boer maak 'n plan* (literally, "the farmer makes a plan," but more metaphorically, an Afrikaner will always find an *ad hoc* solution to the challenges posed by a changing situation). Also, since a very high proportion of the country's white male population served in the military, the country's industries – including the arms sector – were permeated with individuals having continuous, hands-on combat experience, a natural interest in military technology, an ethos of self-sufficiency and a strong educational grounding in the hard sciences and engineering. The arms industry seems to have thrived on what van Rensburg calls a "youthful engineering culture," absorbing large numbers of young engineers and technicians, giving them great creative latitude, and generously funding applied military-related research and development. These characteristics encouraged innovation and improvisation in military technology.[33]

Still another interesting (if obvious) characteristic of the South African defense industry was the relatively small community of its senior leaders, scientists and engineers. For most of the *apartheid* era, the industry drew its top personnel from the white population, and largely from the even smaller subgroup of Afrikaners.[34] This meant that industry insiders, key military leaders and top government officials came from the same limited population, attended many of the same schools and were strongly connected by bonds of ethnicity, family, church, school and social life. This did not attenuate significant cultural differences or eliminate deep feuds and grudges (still the stuff of office chatter in South Africa) but did promote a certain uniformity of world-view and common approach to problems. It also enabled industry, government and military leaders to resort to informal connections to communicate, influence decisions and solve problems. On the other hand, it contributed to a culture of secretive decision-making lacking transparency and accountability.

Two men deserve much of the credit for the evolution of the South African arms industry, for the rapidity of its development, and for the form it ultimately took. These were P. W. Botha, South Africa's longest serving Minister of Defence (1966-1978) and subsequently Chief of State from 1978-1989. The other was "Commandant" Piet Marais, who headed the arms producing parastatal, Armscor, in its most productive years. Botha is widely remembered in the rest of the world as the finger-wagging personification of South African intransigence during the *apartheid* years, an icon of regressive state policies. Yet that caricature does little justice to this complex, capable man.[35] His service as Minister of Defence spanned the formative years of the consolidation of the South African arms industry under Armscor's supervision (and, thus, under the direction of Botha's Department of Defence). Virtually all of the significant achievements of that industry occurred while he was Minister of Defence, Prime Minister or State President. The effective supervision and achievement of synergy across disparate communities of civil servants, strategists, military staffs and private sector industrialists was an incredibly complex and difficult process. According to insiders in the military and arms industries in those years, Botha's strong personality, integrity, genuine leadership qualities, and ability to find and delegate responsibility to capable subordinates were indispensable to the ultimate results.

"Commandant" Piet Marais was the leader and guiding spirit of Armscor during much of the *apartheid* era. He exemplified the organization's idiosyncratic (and heavily Afrikaner) complexion. Marais may have seemed an odd choice to lead the complex, technologically sophisticated and politically fraught arms industry. He was an unpretentious farmer from the Karoo who had completed only a year of agricultural technical training beyond secondary school. His path to the leadership of Armscor lay through Minster of Defence P.W. Botha, a family friend and hunting partner. Botha apparently recognized in the avuncular farmer a genius for practical management. The choice proved inspired.[36] Marais had the insights and interpersonal skills necessary for effective leadership of a community of South African engineers, many of whom shared his rural upbringing. His subordinates remember him as a grandfatherly figure, pipe in hand, always able to smooth, cajole, encourage and reconcile. None seemed to doubt that he cared deeply about the welfare of his people and most seem to credit Marais for establishing much of Armscor's organizational culture and effectiveness. Marais oversaw the remarkable development of his country's arms industry.[37]

Sources of Arms

Despite the sophistication it ultimately achieved, the industry evolved in a somewhat *ad hoc* way, particularly in the early years. Its coherence and capability matured dramatically over the course of about a decade, coinciding directly with Botha's tenure as Minster of Defence. What ultimately emerged from these efforts was a coordinated military procurement system that drew materiel in a well-organized manner from three separate – but interrelated – sources. The most important in the early years continued to be the international market, a source that became increasingly less accessible to South Africa in the wake of the UN arms embargoes in 1963 and 1977. The second was the South African private sector, typically in the form of defense-related subsidiaries of larger industrial or financial conglomerates. The third source was a public sector armaments industry – an idiosyncratic creation of the *apartheid* state – deliberately developed to manufacture the armament unavailable either from foreign suppliers or domestic commercial enterprises. It was this public sector armaments industry that eventually controlled and dominated the South African market, though each of the three sources continued to make important contributions to South Africa's military capabilities throughout the *apartheid* era.

Foreign Sources

In the immediate aftermath of World War II, South Africa depended largely on the United Kingdom for its modest defense needs, acquiring weapons such as *Vampire* jet fighters and *Centurion* main battle tanks.[38] However, when it left the Commonwealth in 1961, South Africa began to shop more widely. Its initial purchases included small arms from Belgium, light jet aircraft from Italy and armored scout vehicles, submarines, combat jets and helicopters from France. (One authority says that France supplied 45 percent of South Africa's arms imports between 1965 and 1969.) Italy apparently also served as a conduit for arms from various other countries, including the United States.[39]

The international community at first barely honored the 1963 UN "voluntary" embargo against supplying arms to South Africa, one author referring to it contemptuously as the "marshmallow" embargo, but the situation changed in the mid 1970s.[40] Worldwide pressure exerted by international anti-*apartheid* activists increasingly isolated South Africa from arms suppliers in the West, a process that culminated in the UN "mandatory" embargo of 1977.[41] Still, while the rest of the world backed away, South Africa embraced several other nations, particularly Israel.[42] At first, South Africa imported Israeli small arms, missile gunboats and anti-ship missiles.[43] Following a 1976 visit to Israel by its Prime Minister, South Africa took delivery of a further range of Israeli military armaments, including high-technology electronic and electro-optical equipment.[44] A close security relationship grew quickly and persisted for about a decade in spite of the "mandatory" UN embargo. Israeli engineers were very active in South Africa's ballistic missile program and in the upgrade of the earlier Mirage 3 combat aircraft to the new *Cheetah* configuration, as well as the manufacture of the new *Cheetah* C, essentially an Israeli *Kfir*.[45] The armaments industries of the two countries appeared to be cultivating a variety of close connections between the late 1970s and early 1980s.[46] However, the association apparently soured in the mid 1980s, when Israel began to distance itself politically from the South Africa, and scaled back its defense relationships.[47]

With the advent of the "mandatory" UN embargo in 1977, the political cost of arms sales to South Africa became sufficiently high to prevent overt arms transfers from most external sources, though by that point the country had achieved a substantial self-sufficiency in arms production.[48] Still, South Africa continued to depend on external sources for some materiel and pursued vigorous programs of clandestine arms (and technology) transfers and intellectual property theft. The South Africans also found military applications for much dual-use technology not subject to embargo, including electronics, computers and vehicle components (such as the robust German bus engines installed in South Africa's new armored combat vehicles).[49] Dual use materiel from abroad significantly facilitated the upgrading of South African military capabilities, as the country was able to redesign, modify and retrofit weapons with more modern technology.[50] By the 1970s, the country's engineers had learned do this very quickly, a unique capacity of South Africa's defense industry that continues to the present day.[51]

Armaments were not the only important security-related resources imported from overseas. The South Africans also managed to recruit and retain a considerable number of foreign engineers and technicians for their defense industries, attracting them with high salaries and comfortable lifestyles. In the 1960s, these included out-of-work British aerospace engineers.[52] By the late 1970s, the community was supplemented with a sizable contingent of Israelis.[53] Then, in the early 1980s, the administration of French President François Mitterand nationalized the French aerospace industry and the South Africans induced senior corporate personnel connected to France's advanced aircraft programs to immigrate to South Africa.[54] The expatriate engineers and technicians from a number of countries remained an important feature of South Africa's armaments industry throughout the *apartheid* era.

While the foreign sources of armament were important, and expatriate technicians played key roles, South Africa proved eminently capable of indigenous weapons design and arms production. These activities were shared in a unique partnership between the public and private sectors.

The South African Private Sector

In the early 1960s, the South Africans revived their World War II model of arms production, and a range of private sector enterprises once again became heavily engaged in defense production, initially under license from abroad. Such enterprises typically were subsidiaries of larger industrial or financial conglomerates.[55] These included, among others, companies that produced small arms,[56] military ground vehicles, vehicular components,[57] and communications equipment.

Electronics firms were some of the first to evolve specifically to meet South African defense needs. Ironically, these got a substantial boost by buying out local subsidiaries of multinationals when UN sanctions obliged the latter to disinvest in South Africa after 1963.[58] Among the accomplishments of the local electronics enterprises was the fielding of frequency-hopping tactical radios by the late 1960s, the first in general military usage. (The technology subsequently was exported to the NATO consumers.) The South African electronics firms also developed extremely sophisticated and effective state-of-the-art signals intelligence and electronic warfare equipment for use by the country's ground forces and aircraft.

As early as the mid 1960s, South Africa's new arms industries were achieving significant rates of production and significant increases in technological capacity, progressing to the manufacture of a light jet.[59] By that time, the growing public sector arms conglomerate, Armscor (and its predecessor organizations) had begun to organize and control these efforts, ultimately absorbing some of them into its own expanding public sector production base.

Private sector industries adapted relatively easily to the military's growing requirement for small arms and military electronic applications. They also adjusted quickly to licensed production of ground combat vehicles, eventually developing the capacity to build innovative new versions. However, South Africa had to look beyond the private sector to assemble, upgrade and produce its increasingly sophisticated aerospace equipment, electro-optical applications and (ultimately) space and nuclear materiel. In 1964 the government provided the financing to set up the Atlas Aircraft Corporation, chartered initially as a private sector enterprise to assemble and produce military aircraft. Within five years, Atlas had been absorbed into Armscor, the growing arms parastatal.[60]

The South African Public Sector

Although the private sector continued to produce arms throughout the *apartheid* years, it was the public sector industry that ultimately organized, coordinated and controlled the production of military materiel. That industry consisted of two major actors: the Council for Science and Industrial Research (CSIR) and the Armaments Corporation of South Africa (Armscor). Each had very different origins and purposes. CSIR was formed immediately after World War II to promote the maturation of science and industry in South Africa, while Armscor coalesced in the 1960s as the organization established to oversee materiel acquisition for the South African military. Of the two, CSIR was the older, but Armscor became by far the most significant in terms of acquisition policy and arms production. The origins of both are an interesting study in South African security decision-making during the *apartheid* years.

Armscor's earliest roots appeared with the new National Party government in 1948 – as an Advisory Committee on Union Defence Equipment Requirements. This was instituted within the Department of Defence. A year later, the Advisory Committee became the Defence Resources Board, now charged with overseeing the development of an indigenous arms production capability in the private sector. Then in 1951, the government chartered the Defence Ordnance Workshop (one of the two government World War II production facilities still intact from the war years) to coordinate the repair and manufacture of the weaponry already on hand.[61] The government also established a Defence Production Office whose primary focus was the acquisition of foreign licenses for local production from foreign arms manufacturers.[62] All of these institutions were subordinated to the Department of Defence.

In the early 1960s, procurement of arms from domestic sources still was very low,[63] but the threat of UN sanctions (first imposed in 1963) added urgency to earlier efforts.[64] The threat of sanctions also persuaded the South Africans to invest in public sector arms production. So the country took a significant step in 1964, combining the Defence Production Office with the Defence Resources Board, and recasting the new entity as the Defence (or "Munitions" or "Armaments") Production Board. Still within the Department of Defence, this new organization now was chartered both to manage military acquisition and oversee a domestic arms industry.[65] By this time, it had acquired two of its own production subsidiaries.[66] In 1966, this Defence Production Board was recast as the Armaments Development and Production Corporation of South Africa and, finally, in 1968 as the Armaments Corporation of South Africa (Armscor), assuming the final name for the organization that became the backbone of the South African arms industry in the *apartheid* years.[67]

As of 1 April 1968, Armscor's role was to procure military materiel for the South African Defence Force from all available sources and produce what could not be obtained from abroad or from the indigenous private sector.[68] By this time, Armscor had evolved into a public sector holding company with a board of directors chaired by the commanding general of the Defence Force. The board itself consisted of leading scientists, financiers and businessmen. The Prime Minister (later, the State President) appointed the members of the board. By 1968, enterprises subordinate to Armscor had grown from the original two to eleven production, test and research facilities serviced by a host of private sector contractors. Virtually all Armscor's acquisition, research, test and production activities were state secrets.[69]

While Armscor had by 1968 become the center of South Africa's arms industry, it still shared the role of arms research with the other public sector entity – the Council for Scientific and Industrial Research (CSIR). The South African Parliament had created CSIR in 1945 and while its primary function was to perform scientific study in support the domestic economy as a whole, the institution seems to have had a small, covert military role almost from its inception, a role that expanded over time. CSIR was an autonomous state agency, unlike Armscor, which was embedded in the Department of Defence. In 1954, CSIR's military-related research was focused into the National Institute for Defence Research (NIDR), a new subordinate entity. The Minister of Defence chaired NIDR, with the president of CSIR as co-chairman. Probably in anticipation of UN sanctions, the CSIR security agenda was significantly augmented after 1962 with the creation of a Defence Research Council. When the United Nations imposed its first arms embargo in 1963, CSIR began to devote substantially more effort to military research and development.[70]

Several CSIR subsidiaries subsequently migrated to Armscor over the years. One of the earliest of these was Mechem, an enterprise that researched mine and countermine technology, ultimately designing mine-protected vehicles and innovative explosives detection systems. Mechem was transferred from CSIR to Armscor in the 1960s.[71]

Still another loosely organized group within CSIR interested in rocketry had by the early 1960s begun to focus on missile technology.[72] This effort grew substantially in the 1960s, resulting in the joint development with French partners of the *Crotale* surface-to-air missile[73] and the production of the first practical pilot helmet sight slaved to an air-to-air missile.[74] By 1978 the CSIR missile group had become so sophisticated that a major new defence company – Kentron – was spun out of it, specializing in sophisticated missile and aerospace technology. Kentron then became a subsidiary of Armscor.[75] Also, as part of an overall military and industry reorganization of defense, much of CSIR's military research capacity was transferred to Armscor in 1978.[76]

Despite the loss of Mechem, Kentron and the research assets, CSIR continued to play a role in military-related research, development and testing – still somewhat independently of Armscor. There were substantial differences in emphasis between the two organizations. CSIR remained a public sector agency more devoted to "pure" science and unconstrained research, in contrast to the much more "applied" engineering focus of Armscor.

CSIR also functioned somewhat as a training agency for the defense industry: it was common for young engineers to find an initial employment in CSIR, continue their education under CSIR sponsorship, then eventually find permanent appointments in Armscor or the private sector.[77] In maintaining CSIR's defense research capabilities separate from Armscor, the South African government probably demonstrated a desire for unconstrained security-related research that maintained some autonomy from the defense establishment, and an entity that performed "purer" science than the more engineering-oriented Armscor. Both organizations – and their separate roles – have continued to the present.[78]

Evolution of Armscor

Reacting to their sobering military experience in Angola in 1975, the South Africans began a massive reorganization of Armscor and its subsidiary institutions in 1976, with the clear intention of radically enhancing their combat capabilities and self-sufficiency in arms.[79] Some features did not change. Armscor remained a subordinate agency of the Department of Defence, represented in key national planning and policymaking fora and accountable only to senior Defence officials. It continued to function as a holding company, managing the efforts of a number of subordinate parastatals and a large network of private sector subcontractors, and still was dependent on private sector financing and components.[80] Like the decisions of the State Security Council that made all of the country's security-related decisions, Armscor's research, development and acquisition efforts remained shrouded in secrecy.[81]

The reorganization after 1976 significantly expanded Armscor's role and output. Its overall research and development programs more than doubled between 1977 and 1981.[82] By the late 1970s, it had become the key actor in a sophisticated military-industrial complex, having, in Gavin Cawthra's words, "assumed sole authority for arms acquisition and military research and development. . . . It also carried out around 80% of domestic production, which increased rapidly after the imposition of the mandatory UN embargo [in November 1977]." Because of its mandate to procure needed armament from any available source, Armscor's officials now became even more heavily involved in covert efforts to procure military technology from abroad in violation of international sanctions.[83]

The South Africans had by this time developed a fairly standard process for developing new military materiel under Armscor's purview. This typically took the form of temporary work teams whose leadership included a uniformed military project officer tasked to protect the equities of the consumer (the South African Defence Force), an Armscor program manager concerned with the programmatic issues of oversight, plans and budgets, and the project manager from a production enterprise contracted to perform the actual work.[84] Under ideal circumstances, such *ad hoc* teams enabled the arms industry to rapidly conceive, design and produce effective battlefield implements, although (perhaps not surprisingly) military and industry insiders still recall bitter disputes between military users and arms industry engineers over the details of fielded materiel.

Armscor supervised all South African military acquisition, but its direct subsidiaries were not the only elements in the evolving defense industry. In the 1960s, close to 1000 South African private sector enterprises produced armaments. By the mid 1980s these had grown to over 2000 firms, and by the end of the 1980s, to 3000. These private sector contractors and subcontractors were like other South African industrial enterprises – typically subsidiaries of large, diversified corporate groups that were themselves owned or controlled by one of the major South African financial-mining-industrial conglomerates.[85]

By 1985, Armscor itself had grown to include twelve production and nine testing/evaluation subsidiaries.[86] Whatever the source, Armscor now coordinated the efforts of a vast industry developed to produce materiel to satisfy South Africa's escalating need for arms.[87] An industry official asserted in 1983 that the country was producing 74 percent of all its war materiel. By 1985, almost 100 percent of the materiel purchased by the ground forces was locally made.[88]

New Military Materiel

The arms acquired by South Africa in the early 1960s had been relatively simple – small arms and crew served weapons, scout cars and jet trainers. However, the sophistication of the weaponry – and South Africa's capacity to produce it – escalated over the decade. By the mid 1960s, the South Africans were acquiring new naval and military air capabilities, purchasing frigates from the United Kingdom and submarines from France.[89] By 1971, Armscor's new Atlas Aircraft Corporation was on the verge of producing the French Mirage-3 fighter-bomber and the more advanced Mirage F-1 interceptor.[90] By this point Atlas was assembling the Mirage-3 under license, was manufacturing turbojet engines for a less sophisticated light jet trainer and was producing an increasing quantity of other aerospace equipment.[91]

The mid 1970s saw a further dramatic increase in indigenous arms production, initially under license. This included local assembly of French-designed *Mirage* F1 jet as well as the French-designed *Alouette* III and *Puma* helicopters.[92] By now South Africa also had begun to field significant product improvements to many previously licensed weapons, and began to build indigenous designs specifically suited to its operational needs. Among the latter was a new, wheeled infantry fighting vehicle fielded in 1975 (the *Ratel*), a vehicle that soon appeared in a variety of configurations – fighting, logistic, command, troop carrier and mortar firing platform – and saw almost immediate combat employment in Angola.[93] The country also began to produce several lines of transport, combat support and fighting vehicles designed to defeat land mines – a growing regional hazard.[94] The new weaponry had significantly enhanced South Africa's combat capabilities by the late 1970s.

The years after 1977 saw a steady stream of even more sophisticated military materiel from South African producers. Much of this was ingenious reformulation of older technology, but by the early 1980s South Africa had succeeded in building some weaponry that genuinely could be characterized as "cutting edge." Even here, much of the technology actually originated elsewhere, but the South Africans were able to combine ideas, processes and materiel in innovative new ways. Their accomplishments were noticed by other military establishments: as soon as they began selling arms abroad in the early 1980s, their materiel was eagerly purchased by the competing parties in the Iran-Iraq war – the world's highest technology conflict of that era – and it played a significant role for both sides.[95]

By the late 1970s, the growing inventory of new (or enhanced older) arms in the hands of South Africa's military forces included an extensive range of ground, air and naval equipment, with the ground forces receiving the greatest assortment. A new infantry rifle essentially copied the Israeli *Galil* assault rifle.[96] The South Africans upgraded and modernized their older British-designed *Centurion* tanks, and continued their development of mine-resistant transport vehicles.[97] Within another decade, they had also developed and fielded a new, wheeled armored reconnaissance vehicle (the *Rooikat*), possibly the best of its type in the world.[98] But their most visible successes were in field artillery. First, they introduced a new self-propelled multiple rocket launcher, inspired by a Soviet model widely used in Angola, yet with superior mobility, lethality and range, employing it with devastating effect against the Cubans and Angolans.[99]

Then they fielded the long-range G-5 and G-6 155mm guns, possibly the world's best 155mm systems at the time of introduction in the early 1980s.[100] Shortly after being fielded, these guns also were combat tested in Angola, with stunning success. By 1985, the South Africans had unambiguously regained the regional edge in artillery capability.[101] The G-5 proved a popular export despite the UN embargo.

The capabilities of the South African Air Force and Navy also were enhanced between the mid 1970s and mid 1980s by a series of upgrades. These included the retrofit of the older French-designed medium-lift *Puma* utility helicopter into a very successful upgrade locally called the *Oryx*, still a world leader for high ambient temperature high-altitude ("hot and high") operation.[102] Another program (with substantial Israeli assistance) upgraded the French-designed *Mirage* 3 jet fighter into a modernized version locally called the *Cheetah*.[103] Nor was the South African Navy entirely neglected. Its three UK-built *President* Class frigates had been modernized earlier,[104] and South Africa now built a new military replenishment ship from scratch.[105] In the wake of the Prime Minister's visit to Israel in 1976, the country also purchased light ocean-going missile boats and patrol boats from Israel, along with a state-of-the-art Israel anti-ship missile, and the rights to produce all of these locally. The new Israel boats were delivered in 1977 and South African-built versions were completed by 1982.[106]

The Israeli connection after 1976 seems to have energized some other high-tech South African programs as well, especially those with electronic, electro-optical, space and aviation applications (including the aforementioned upgrade of Mirage 3 into the *Cheetah*). South Africa began a collaborative effort with Israel to produce a ballistic missile based on Israel's *Jericho* II, purportedly for launching low-earth orbit satellites (but covertly intended as a delivery system for nuclear weapons). Israeli experience, research and technology probably also contributed to South African development of armored vehicle and protective armor applications in the 1970s and 1980s.[107]

By the late 1970s, South Africa's success in building ordnance delivery systems was matched by the growing sophistication of its projectile ordnance. This ranged across the spectrum of guided and unguided projectiles. The Armscor subsidiary Kentron provided tactical guided missiles for air-to-air, ground-to-air, anti-armor and anti-ship usage.[108] The South Africans quickly exploited the sophisticated Soviet missile technology like the SA-8 their forces captured in Angola in the 1980s.[109] In general, their guided missile developments kept pace with all but the most advanced developments in the West, and in several categories, they achieved "leading-edge" technological leadership. The latter category included a progressive evolution of "smart" munitions starting with a TV-equipped glide bomb and progressing to laser-designated guided devices.[110] They included air-to-air missiles slaved to pilot helmet sights, described earlier.

The South Africans also began to apply innovative new technology to their unguided projectiles, including base-bleed artillery munitions for enhanced range and clever new fragmentation designs to increase lethality.[111] For its high-performance ground attack aircraft, South Africa developed fragmentation bombs (120 and 250 kilogram weights) with a variety of applications. These were tested in combat in Angola, refined, and used with considerable success against personnel and vehicles throughout the 1980s.[112] The country also energized the domestic production of explosives and propellants for its ordnance. In 1970 Armscor had taken over two enterprises founded to research and produce these materials for military munitions.[113] In addition to standard propellants, one of these (Somchem) now began to build missile propulsion systems.[114]

South African military production was not limited to transport, ordnance and delivery systems. Its industries provided a growing inventory of intelligence collection and electronic warfare materiel, with both ground and air applications.[115] These too probably reflected substantial Israeli contributions but nonetheless were genuinely "state-of-the-art," reflecting an expertise that the South Africans have retained to the present. By the early 1980s, the Armscor subsidiary Kentron had designed and fielded a state-of-the-art remotely piloted vehicle (RPV), locally called the *Seeker*, a propeller-driven craft with a 40 kilogram payload and a 200 kilometer range. This system provided real-time intelligence and targeting video imagery to a ground station.[116]

South Africa's interests extended to strategic weaponry and weapons of mass destruction. By the late 1980s, it had an active program of chemical and biological warfare research, and in 1982 produced the first of a small number of low-yield tactical nuclear weapons. (The latter were designed for delivery by combat aircraft, but ultimately intended for delivery by ballistic missile as well.[117]) South Africa had configured several Boeing 707 transport aircraft for electronic intelligence (ELINT) and military aerial refueling, so its old British-made *Buccaneer* tactical bombers conceivably could have had a very long "strategic" reach.[118] South Africa also had invested in a space program, progressing to the testing of booster devices and the assembly of satellites.[119]

A Mature Arms Industry for the Apartheid State

At the end of the *apartheid* era, the South African military establishment had an astonishing amount of state-of-the-art materiel developed exclusively for its unique requirements and produced by its own arms enterprises. The UN embargos had encouraged the development of an industry possessing in Heitman's words, "extraordinary depth and breadth of capability"[120] although another analyst characterizes the industry more as a "high technology combat arsenal" suited to immediate tactical needs than an industry suited to long-term strategic research and production.[121] Still, South Africa's investment in arms "did bring the country a useful measure of strategic independence and the armed forces a range of equipment optimized for their operational style and demands of the theatre."[122]

The economic importance of the arms industry to the country had become as significant as its role in providing weaponry to the security forces. In 1989 that industry accounted for fully nine percent of South Africa's manufacturing manpower. By this point, 3000 private sector enterprises serviced some aspect of the arms industry as contractors, subcontractors or suppliers.[123] Armscor alone was the fifteenth largest employer in the country.[124] Arms production comprised 6.6 percent of all South African manufacturing output and 1.5 percent of the country's gross domestic product.[125]

In the reorganization after 1976, Armscor had been chartered to control South African arms imports and exports, and in 1982 it created a new subsidiary – Nimrod –with responsibility for marketing South Africa's arms abroad.[126] Nimrod made its first major international appearance in that same year at an international arms exhibition in Greece, and arms assumed an increasing role in South African exports from that point on.[127] In 1984 the UN Security Council grudgingly acknowledged South Africa's growing success by passing a resolution embargoing the purchase of South African arms,[128] a resolution widely ignored by the world's military consumers.[129] But despite the increasing importance of exports, Armscor's primary focus still remained the needs of the South African Defence Force.

Although it enjoyed great success in fielding state-of-the-art materiel, South Africa's expanding armaments production in the late 1970s ultimately brought significant problems to the country as well. Investments in arms had not been subjected to public referendum and required an enormous diversion of resources from other sectors. Funding from exports could not buffer the armaments industry from a variety of problems. By the early 1980s, the industry had begun to experience problems of excess capacity and declining local demand, and Armscor was obliged to lay off workers after the early 1980s (some 3000 personnel between 1981 and 1985). By 1984, Armcor's financial problems had resulted in a significant government investigation.[130]

One major problem was the limited South African market, requiring extensive retooling for relatively short arms production runs. And international sanctions were beginning to take a toll. Even though the country had begun to make concerted efforts after the early 1980s to finance its arms by selling military hardware abroad, the limited demand (the one reliable client being the South African government) prevented the industry from achieving any great advantage in economies of scale. So much research and production had been devoted to defense that the core arms producers simply did not diversify into products for the civilian market prior to the advent of majority rule.[131] Then, the country's isolation from the rest of the world restricted its access to new ideas and technology developed elsewhere. Nor was the defense industry subject to the salutary pressures of local competition. South Africa's investment in armaments contributed to a protracted recession in the 1980s.[132] Ironically, by the end of the 1980s, the financial problems caused by South Africa's investment in arms may actually have contributed to the willingness of its white population to accept political change.[133] The buffeted business community had become a powerful advocate for reform, including even the Afrikaner industrialists.

Still, the arms industry had provided the country's leadership with sufficient military capability to forestall political change for years, and when substantive reform did come after 1989, the demise of Soviet power had significantly reduced the value of Soviet patronage to the anti-*apartheid* groups. The clout of the South African *apartheid* state was sufficient to oblige its opponents to negotiate rather than dictate the details of a new South Africa. Whatever the cost, the arms industry had played a role in maintaining state power until the South African government could settle with its opponents without the concurrent pressures of the Cold War.[134]

The end of the Cold War was a watershed event in southern Africa as elsewhere, and the ability of the *apartheid* state to forestall political change until that point must be acknowledged as an achievement. Regional Soviet clients felt the effects almost immediately. The South African Communist Party had been a key political actor in the country since the 1930s, and had very close links to Moscow. The East Bloc and its radical allies were significant patrons for many of the opponents of the *apartheid* state. The anti-South African insurgents in Southwest Africa received substantial backing from East Bloc sources and safe haven in Marxist Angola after 1976. The regime in Luanda had been heavily supported by materiel and military personnel from the Soviet Union, East Germany and other European communist countries, and sustained by up to 50,000 Cuban troops. When the Soviet Union began to unravel in the late 1980s, it proved increasingly unable and unwilling to support these various clients. That weakness directly impacted southern Africa, providing political "space" in South Africa for a negotiated settlement between the government and its internal and external opponents.[135]

In South Africa itself, the advent of the F.W. de Klerk government in 1989 spelled an imminent end to the old order. De Klerk wasted little time in withdrawing from external conflicts[136] and scaling down the defense sector.[137] He released the imprisoned Nelson Mandela in 1990, lifted the ban on the opposition African National Congress (ANC) and accelerated substantive negotiations with domestic opposition groups.[138] De Klerk and Mandela then jointly supervised a Convention for a Democratic South Africa (CODESA), begun in 1991, leading to the adoption of new "interim" constitution in 1993 that formally dismantled the structures of institutionalized racism.[139] (In 1993, De Klerk and Mandela shared a Nobel Peace Prize for their achievements.) A multiracial national election inaugurated majority rule under Mandela's ANC-led interim government in April 1994.

In the process of political reform that ushered in majority rule, the future of South Africa's military establishment and arms industry were by no means foreordained. In the minds of many citizens both were associated with the most unpleasant features of the *apartheid* era and South Africa's population was not initially convinced of their desirability. A small group of intellectuals in the African National Congress had coalesced to study and recommend new approaches to security, and they ultimately would endorse both a conventional military and a supporting industry, but their vision was simply one of many competing for attention in the new South Africa in the early 1990s. The trajectory of the subsequent transition could not have been easily foreseen, and the ultimate results were as surprising as they were interesting. These depended on an unusual conjunction of actors, interests and fortuitous circumstances. It is to these that the story now turns.

Notes

1. Signe Landgren, *Embargo Disimplemented: South Africa's Military Industry* (New York: Oxford University Press, 1989), 37-40.

2. This was due in large part to the mining industry, prominent in South Africa since the late 19th Century, which required a steady supply of engineers and scientists and subsidized South African higher education in a variety of ways. Author's interviews with Arthur J.R. Oates, Director of Business Development, Denel, 12 June 2003; and Brigadier Tony Savides (South African Army, retired), at the time Director of International Marketing, Alvis-OMC (now BAE Land Systems OMC), 18 June 2003.

3. Landgren, *Embargo Disimplemented*, 37-8.

4. The National Party was rooted in the Afrikaner community – comprising approximately 55% of the population of European descent. The Afrikaners traced their roots to the Dutch, German and French settlers lured to southern Africa by the Dutch East India Company settlement after 1652. By the 19th Century, they had developed a heritage of frontier independence, cultural distinctiveness and sense of unique destiny in southern Africa. By the early 20th Century many also carried strong resentment against the United Kingdom due to British colonial aggrandizements, culminating in the Boer War (1899-1902) in which British arms subjugated formerly independent Afrikaner republics. British counter-insurgency measures in that conflict included atrocities that impoverished rural families and left a legacy of great bitterness. The National Party had seen its agenda harden and its chauvinistic inclinations increase over the course of the 20th Century. Many *Afrikaans*-speaking white citizens now sought through the National Party to establish their political space, reaffirm their cultural values and guarantee their economic security. One of National Party's most consistent programs was a form of "affirmative action," – guaranteeing jobs for whites, particularly Afrikaners. For useful details of South Africa political history, see, *inter alia*, William H. Vatcher, *White Laager: The Rise of Afrikaner Nationalism* (New York: Praeger, 1965) and Hermann Giliomee, *The Afrikaners: Biography of a People (Reconsiderations in Southern African History)* (Richmond, VA: University Press of Virginia, 2003).

5. The National Party leadership at the time probably believed that its determination not to "fight Britain's wars," combined with South Africa's remoteness from likely European battlefields, meant that the country faced little military threat. This attitude changed significantly only in the 1960s, reaching its culmination with the South African intervention in Angola in 1975. Helmoed-Römer Heitman, South African correspondent for *Jane's Defence Weekly*, in a personal communication with the author, 12 June 2004.

6. Landgren, *Embargo Disimplemented*, 37-8. The only South African munitions factories not dismantled after World War II were the Defence Ordnance Workshop (which evolved into Lyttleton Engineering Works – *Lyttleton Ingeneurwese Werke* – generally known by its current acronym LIW), the SA Mint .303 Ammunition Factory

and Pretoria Metal Pressings (PMP) a commercial manufacturer of small arms ammunition.

7. This was the Alvor Accord named for the Portuguese town in which the negotiations took place.

8. The South Africans intervened on behalf of the Union for the Total Independence of Angola (UNITA), led by Jonas Savimbi. UNITA at one time had been the most politically radical of the three competing Angolan groups, having established ties to the People's Republic of China, but after 1975 was increasingly beholden to the South Africans. When the South Africans withdrew from Angola in 1976, Savimbi retreated to the remote southeastern corner of Angola where, with continuing South African assistance, he rebuilt his forces and waged an increasingly successful insurgency against the Marxist government in Luanda. In 1975 and early 1976, the United States supported the Front for the National Liberation of Angola (FNLA), headed by Holden Roberto and backed by Mobutu Sese Seko, leader of neighboring Zaïre. US support was organized by the Central Intelligence Agency and funneled through the significant US presence in Zaïre. The FNLA's battlefield incompetence enabled the Soviet-backed MPLA to crush it decisively by early 1976. Zaïre's Mobutu then committed elements of his own army to the conflict in Angola, only to see them routed as well. For an interesting, if somewhat sensationalized, account of the US involvement, see John Stockwell, *In Search of Enemies: A CIA Story* (New York: W.W. Norton & Co., Inc, 1984).

9. This was the Popular Movement for the Liberation of Angola, better known by its Portuguese acronym, MPLA. This faction won the initial phases of the civil war and established itself as the government of Angola in 1976. For an interesting description of this faction and its Soviet and Cuban patrons, see Piero Gleijeses, *Conflicting Missions: Havana, Washington, Pretoria* (Alberton, South Africa: Galago Books, 2003), 300-46.

10. The motives for South African involvement in the Angolan conflict have been analyzed elsewhere and need not be repeated in detail here, but Angola's civil war was part of the violent decolonization of the region, during which South Africa sought desperately to secure itself against the pressures of national liberation. For a brief useful overview, see Fred Bridgland, *The War for Africa: Twelve Months that Transformed a Continent* (Gibraltar: Ashanti Publishing Limited, 1990). For a discussion of the rationale for this intervention, in the words of those responsible for it, see Hilton Hamann, *Days of the Generals* (Cape Town: Zebra Press, 2001), 14-45. For an interesting discussion of these events from the opposite (Cuban) side, see Piero Gleijeses, *Conflicting Missions,* 300-46.

11. See *inter alia*, Bridgland, *The War for Africa*, 12; Hamann, *Days of the Generals*, 21-45; and Helmoed-Römer Heitman, *The South African War Machine* (Novato, CA: Presidio Press, 1985), 170-4.

12. This was as a result of the Clark Amendment, a tangible expression of US Congressional disapproval of the Ford Administration's foreign military adventures. The Vietnam War, a commitment that had traumatized American society, had ended in 1975 with an overwhelming North Vietnamese victory against America's South Vietnamese clients. The US Congress wanted no repeat of such disasters. For a brief, useful discussion of the US change of policy, see Bridgland, *The War for Africa*, 3-11.

13. The insurgents were members of the People's Liberation Army of Namibia (PLAN), the armed wing of the South West African People's Organization (SWAPO). The objective of their insurgency was an independent Southwest Africa (now Namibia) a League of Nations Trust Territory occupied by South Africa since 1915. After 1976 the

Angolan government provided PLAN with bases in Angola, along with training, materiel and other assistance. This enabled PLAN to infiltrate and conduct attacks in northern Southwest Africa.

14. For details of South African military operations in Angola, see Bridgland, *The War for Africa*; Heitman, *The South African War Machine*; Helmoed-Römer Heitman, *War in Angola* (Gibraltar: Ashanti Publishing, Limited, 1990); Willem Steenkamp, *Borderstrike! South Africa into Angola* (Durban: Butterworths, 1983).

15. Scott Thompson, "South Africa and the 1988 Agreements," in *Disengagment from Southwest Africa: The Prospects for Peace in Angola and Namibia*, ed. O. Kahn (New Brunswick, ME: Transaction Publishers, 1991), 117-30.

16. See, *inter alia*, Robert S. Jaster, "South African Defense Strategy and the Growing Influence of the Military," in *Arms and the African*, ed. W. Foltz and H. Bienen (New Haven, CT: York University Press, 1985), 121-52; and Greg Mills, *South Africa: The Total National Strategy and Region Policy During the Botha Years, 1978-1989*. Doctoral Dissertation, University of Lancaster, UK, 1990.

17. This process was reflected in the South African State Security Council, established by statute in 1972, and its implementing organ, the National Security Management System. The implementing legislation was the Security, Intelligence and State Security Act of 1972. Standing membership of the committee included the Prime Minister (later, State President), ministers of Defence, Police, Foreign Affairs and Justice, and senior officials of the Defence Force, Police, and Security Intelligence among others. See, *inter alia*, A. Seegers, "South Africa's national security management system, 1972-1990," *Journal of Modern African Studies* 29, no. 2 (1991): 253-273; and K. Grundy, *The Militarisation of South African Politics* (London: I. B. Tauris, 1986).

18. See, inter alia, Chris Alden, *Apartheid's Last Stand: The Rise and Fall of the South African Security State* (New York: St. Martin's Press, 1996); Deon Geldenhuys, *The Diplomacy of Isolation: South African Foreign Policy Making* (New York: St. Martin's Press, 1984); K. Grundy, *The Militarisation of South African Politics*; and Greg Mills, *South Africa*.

19. Protest riots had begun in the black township of Soweto just outside of Johannesburg and quickly spread throughout the rest of the country. This was the beginning of a long period of unrest in South Africa that would profoundly mark an entire generation.

20. The new strategy, in effect from 1978 to 1989, was associated preeminently with the leadership of Prime Minister (later State President) P.W. Botha, advanced to the office of Prime Minister in September 1978. The strategy itself was first officially articulated in a 1977 *White Paper on Defence*. See, *inter alia*, Robert Davies and Dan O'Meara, "Total Strategy in Southern Africa – An Analysis of South African Regional Policy Since 1978" in *Exporting Apartheid: Foreign Policies in Southern Africa 1978-1988*, ed. S. Chan (New York: St. Martin's Press, 1990), 179-217; K. Grundy, *The Militarisation of South African Politics*; G. Mills and S. Baynham, "South African Foreign Policy, 1945-1990," in *From Pariah to Participant*, ed. G. Mills (Johannesburg: South African Institute of International Affairs, 1994), 18-33; and D. Sole, "South African Foreign Policy Assumptions and Objectives from Herzog to De Klerk," *South African Journal of International Affairs* 2, no. 1 (1994): 1.

21. J. Cock and L. Nathan, "Introduction" in *War and Society: The Militarisation of South Africa*, ed. J. Cock and L. Nathan (Cape Town: David Philip, 1989), xiii.

22. The sanctions were imposed by UN Security Council Resolution 418.

23. See, *inter alia*, J. Hanlon, *Apartheid's Second Front: South Africa's War Against*

its Neighbours (London: Penguin, 1986); William Minter, *Apartheid's Contras· An inquiry into the roots of war in Angola and Mozambique* (Johannesburg: Witwatersrand University Press, 1994); Robert M. Price, "Pretoria's Southern African Strategy," in *Exporting Apartheid: Foreign Policies in Southern Africa 1978-1988*, ed. S. Chan (New York: St. Martin's Press, 1990), 145-69; and Willem Steenkamp, *Borderstrike! South Africa into Angola* (Durban: Butterworths, 1983).

24. South Africa's nuclear strategy, though never publicly articulated, was much more nuanced than merely the threat of nuclear strikes against regional opponents. The South Africans apparently believed that a likelihood of their nuclear use in a regional conflict would impel the United States to intervene on their behalf. Helen E. Purkitt and Steven F. Burgess, *South Africa's Weapons of Mass Destruction* (Bloomington, IN: Indiana University Press, 2005), 79-84; Hannes Steyn, Richard van der Walt and Jan van Loggerenburg, *Armament and Disarmament: South Africa's Nuclear Weapons Experience* (Pretoria: Network Publishers, 2003), 43, 61-71.

25. Annette Seegers, "Apartheid's military: its origins and development," in *The State of Apartheid*, ed. W. James (Boulder, CO: Lynne Reinner, 1987).

26. In 1965, the white minority regime of Ian Smith issued a unilateral declaration of independence from the United Kingdom. By the late 1960s, Rhodesia was facing an escalating insurgency conducted by two indigenous liberation armies. The mobile, highly motivated forces of Rhodesia's white minority government generally were able to best the insurgents in direct confrontations, so the latter resorted to asymmetric attack, placing considerable emphasis on mines and booby traps. This forced the Rhodesians to experiment with air mobility and mine-resistant equipment. Beginning in 1967, the South Africans assisted the Rhodesians in a limited and semi-covert way. They subsequently borrowed ideas about airmobile warfare and mine-resistant vehicle design from the Rhodesians, though in both cases South African adaptations became significantly more sophisticated than the original prototypes. See, *inter alia*, Peter Batchelor and Susan Willett, *Disarmament and Defence: Industrial Adjustment in South Africa* (Oxford: Oxford University Press, 1998), 31; H. Ellert, *The Rhodesian Front War* (Gweru, Zimbabwe: Mambo Press, 1993), 110-23.

27. See James P. McWilliams, *Armscor: South Africa's Arms Merchant* (London: Brasseys, 1989), 54-5; and Landgren, *Embargo Disimplemented*, 218-23. To date, the full extent of the South Africa-Israel relationships remains undocumented, though South Africans involved in these programs now are increasingly willing to talk about them. The aircraft industry and space program seem to have been key areas of bilateral cooperation. South African officers also have told the author of extensive (though covert) Israel training programs in ground force tactics and equipment, so it is likely that South African electronic warfare and armored equipment technology also bears some Israeli antecedents.

28. Oates interview, 12 June 2003; Savides interview, 18 June 2003.

29. Batchelor and Willett, *Disarmament and Defence*, 26.

30. Beginning in the 1970s, South Africa recruited and deployed several effective battalion-sized non-white military units (typically officered mainly by whites), though the total numbers were small. See, for instance, Kenneth W. Grundy, *Soldiers without Politics: Blacks in the South African Armed Forces* (Berkeley, CA: University of California Press, 1983); and Jan Breytenbach, *The Buffalo Soldiers* (Alberton, South Africa: Galago, 2002). Heitman also calls attention to the military forces organized and controlled by the South Africans in Southwest Africa (now independent Namibia). Most of the South West

Africa Territorial Force (SWATF) was comprised of locally recruited non-whites, and by the late 1980s, was officered by non-whites up to the company level. The Southwest Africa police counterinsurgency unit was almost entirely non-white as was the home guard. Heitman says that the Vorster Administration had firmly settled the issue of non-white officers in the South African Armed Forces: they were to enjoy the same perquisites and authority as their white colleagues. Heitman, private communications with the author, 12 June 2004 and 20 April 2005.

31. See, for instance, Gedahlia Braun, "South Africa Under Black Rule (Part 1)," *American Renaissance*, 9, no.7 (July 1988) http://www.amren.com/987issue/987issue.html, (10 January 2005). Within South Africa's white population, the *swart gevaar* ultimately combined an almost inchoate dread of communist armies from abroad with loathing of communist infiltrators within (whose activities all were orchestrated, presumably, by the Kremlin).

32. This was evident in the continued popularity of Boer War memoirs and in the popular literature of the era. See, *inter alia*, Paul L. Moorcraft, *Africa's Superpower* (Johannesburg: Sygma Books, 1981); Willem Steenkamp, *The Soldiers* (Cape Town: Don Nelson, 1978); Siegfried Stander, *Like the Wind: The Story of the South African Army* (Cape Town: Saayman & Weber, 1985); and Heitman, *The South African War Machine*, 1985. For a contrary view that nonetheless reaffirmed the capabilities of the South African Defence Force, see Gavin Cawthra, *Brutal Force: The Apartheid War Machine* (London: International Defence and Aid Fund for Southern Africa, 1986).

33. Oates interview, 12 June 2003; Savides interview, 18 June 2003; author's interview with Alan Holloway, Vice Chairman of AMD and Managing Director of ANSYS Integrated Systems, 21 June 2004. Information also derived from Johan van Rensburg, Commercial Specialist, US Commercial Service, in a private communication with the author, 11 June 2004.

34. Batchelor and Willett, *Disarmament and Defence*, 26, 202.

35. Botha's Administration deserves credit for initiating the processes of political reform that came to fruition in the subsequent de Klerk Administration. Despite South Africa's adoption of the "Total Strategy" – or perhaps even because of it – Botha's tenure as chief of state was marked by a variety of South African efforts to find a formula for political reform that would not at the same time overwhelm the white minority culturally or politically. However, an increasingly confident domestic opposition with growing international support inevitably undermined these efforts. The close connection between the opposition African National Congress (ANC) and the South African Communist Party, along with the locally perceived linkages between the ANC and the Soviet intelligence services, inevitably raised profound suspicions within South Africa's economic and governing elites about the ANC vision for a future South Africa. For detail, see William Cobbett, Daryl Glaser, Doug Hindson and Mark Swilling, "A Critical Analysis of the South African State's Reform Strategies in the 1980s," in *State, Resistance and Change in South Africa*, ed. P. Frankel, N. Pines and M. Swilling (London: Croom Helm, 1988), 19-51; and Steven Friedman, Editor, *The Long Journey: South Africa's Quest for a Negotiated Settlement* (Johannesburg: Ravan Press, 1993).

36. Author's interview with Professor Realize Marais Ferreira, University of South Africa, a close relative of Piet Marais. Marais' military title of "Commandant" (Major) derived from his position in his local community Commando (militia).

37. Author's interviews with Alan Holloway and Hannes Steyn, 11 June 2005. See also Hannes Steyn, Richardt van der Walt and Jan van Loggerenberg, *Armament and*

Disarmament: South Africa's Nuclear Weapons Experience (Pretoria: Network Publishers, 2003), 49.

38. Landgren, *Embargo Disimplemented*, 39.

39. Ibid., 42, 215. Acquisitions from France included *Daphne* class submarines, Mirage 3 combat jets, *Allouette* III, *Super Frelon* and *Puma* helicopters. The Mirage 3 variants included Mirage 3C fighter-interceptors and Mirage 3E fighter-bombers.

40. McWilliams, *Armscor*, 21. See also M. T. Klare, *US Arms Deliveries to South Africa* (Washington: Transnational Institute, 1977).

41. This embargo, too, was very porous, but it did have a significant impact. From 1977 on, the South Africans increasingly were obliged to rely on front companies, covert middlemen and unsavory international arms traders for their supply of foreign arms. See, *inter alia*, Sean Gervasi's testimony before the US House of Representatives in 1977, summarized in *U.S. Arms Transfers to South Africa in Violation of the United Nations Voluntary Arms Embargo, 1963-1977* (New York: UN Centre Against Apartheid, September 1978).

42. For analytical explorations of the motivations behind this relationship, see, *inter alia*, John Adams, *The Unnatural Alliance* (London: Quartet Books, 1984); Benjamin Beit-Hallahmi, *The Israeli Connection: Who Israel Arms and Why* (New York: Phantom Books, 1987); and Deon Geldenhuys, *Isolated States* (Cambridge: Cambridge University Press, 1991).

43. These missile boats were Israel's *Reshef*-class ships, subsequently produced in South Africa under license as the *Minister* class. The missile was the Israeli *Gabriel*, also subsequently produced under license in South Africa as the *Skerpioen* ("Scorpion"). See Landgren, *Embargo Disemplemented*, 219-23.

44. McWilliams, *Armscor*, 54-5.

45. Landgren, *Embargo Disemplemented*, 221; Heitman, personal communication with the author, 12 June 2004 and 20 April 2005. Heitman says that the *Cheetah C* was an Israeli-build *Kfir* equipped with the Atar 09K50 engine designed for the Mirage F-1.

46. McWilliams, *Armscor*, 54-5.

47. Landgren, *Embargo Disimplemented*, 221-3.

48. M. Hough, "The UN arms embargo against South Africa: an assessment," *Strategic Review for Southern Africa* 10, no. 2 (1988): 22-38.

49. Savides interview, 18 June 2003.

50. Landgren, *Embargo Disimplemented*, 192.

51. Batchelor and Willet observe that the South Africans endeavored not to waste time reproducing the research and development already performed elsewhere, but instead emphasized the "capacity for upgrading, add-on and add-up engineering." They reserved their R&D efforts for technology not duplicated abroad, in applications such as mine-defeating vehicle design. Batchelor and Willett, *Disarmament and Defence*, 34. See also *White Paper on South African Defence Related Industries*, Chapter 2, Paragraph 16.

52 McWilliams, *Armscor*, 11-2.

53. Ibid., 54-5.

54. These included the senior Mirage 2000 test pilot, Jean-Marc Pisano, who ultimately became a key executive in the South African aerospace industry. The South African government provided French expatriates with the resources to set up the highly sophisticated, commercially successful private sector enterprise now known as Advanced Technologies and Engineering (ATE). Holloway interview, 21 June 2004; author's interview with Helmoed-Römer Heitman, South African correspondent for *Jane's Defence*

Weekly, 23 June 2004.

55. Key among them were Reunert, Dorbyl/Sandock-Austral and Truckmasters. These, along with OMC, became the enterprises that produced (and upgraded) most of the ground combat and transport vehicles for the South African military and police in the *apartheid* years. After the advent of majority rule in 1994, Reunert shed its vehicle-manufacturing subsidiaries. As this is written in early 2005, virtually all of South Africa's remaining military vehicle manufacturing and refurbishment enterprises have been consolidated in BAE Land Systems OMC. David Botha, *South Africa's Defence Industry: Charting a New Course?* (Pretoria: ISS Occasional Paper 78, September 2003); Savides interview, 18 June 2003; http://www.baelandsystemsomc.co.za, (25 March 2005).

56. The small arms included the Belgian (Fabrique Nationale) FAL assault rifle and the MAG medium machinegun, both of which were state-of-the-art infantry small arms at the time. See Helmoed-Römer Heitman, *South African War Machine* (Novato, CA: Presidio Press, 1985), 38. See also Landgren, *Embargo Disimplemented*, 124, 129-31. The licenses were obtained in 1960 and 1961 from the Belgian armsmaker, though revoked in 1963 with the advent of the first sanctions regime. South Africa ignored the revocation: with minor modifications, South Africa's Lyttleton Engineering (LIW), an Armscor subsidiary, produced the FN rifle (locally designated the R-1) along with the FN MAG machinegun. Production had commenced by 1964.

57. In 1961, South Africa had imported a quantity of surplus French AML *Panhard* wheeled scout vehicles, acquiring the French license to manufacture them. Local production began in 1966. The South Africans renamed the vehicle the *Eland*, upgrading and modifying it in seven distinct models before production ceased in the early 1980s. Heitman, *South African War Machine*, 38. See also Landgren, *Embargo Disimplemented*, 82-3 and McWilliams, *Armscor*, 52-3. The *Eland* was produced in South Africa by Sandock-Austral, a private sector enterprise with roots in mining equipment and dockyard engineering. The government was obliged to finance the creation of a new private sector enterprise to refurbish its 1950s era British-designed tanks. These were British *Centurions*. South Africa had procured its initial lot of 200 *Centurion* tanks from the United Kingdom between 1955 and 1959. In the 1960s it had sold off much of the *Centurion* fleet, so now had to scour the world for additional *Centurions* to complete the inventory the Defence Force desired. The South Africans thoroughly rehabilitated the older *Centurions* with newer guns, upgraded fire control and modernized diesel engines, calling the new version the *Olifant* ("Elephant"). The company created to perform the upgrade was named the Olifant Manufacturing Company (OMC). It remained a private sector enterprise, repeatedly absorbed and divested by larger industrial conglomerates until coming to rest (in 2004) in the BAE Land Systems OMC structure. Savides interview, 18 June 2003. See also http://www.baesystemsomc.co.za, (4 April 2005); and Landgren, *Embargo Disimplemented*, 39, 85-7.

58. Among the South African beneficiaries were Allied Technologies (Altech) whose holding company bought out the shares of ITT, the French STC/Tactel which sold locally manufactured versions of the French Thomson-CSF radios, and (preeminently) Grinaker Electronics (Grinel) that bought out the (UK) Racal franchise. Each of these firms supplied electronics equipment to the South African Defence Forces and also engaged in research and development for military purposes. Landgren, *Embargo Disimplemented*, 140-1; Author's interview with Paul Gerber, Grintek Group Manager for Business Co-ordination and Counter Trade, 13 June 2003.

59. By 1964, the South Africans had begun to produce the Aermacchi MB-326 Ital-

ian-designed jet trainer, locally renamed the *Impala,* an aerial platform used by South Africa in a counterinsurgency/ground attack role to this day. There actually were several distinct versions of the South African *Impala*. The first order (66 aircraft) included some 16 completed aircraft from Italy, the rest consisting of Italian-made components assembled in South Africa. This initial version was the MB-326M (*Impala*-1), the licensing agreement finalized in 1963. A newer version, the MB-326K (*Impala*-2) was licensed by Aermacchi in 1973 for local production. A third variation, the *Impala*-3 was in production in South Africa by the early 1980s. The *Impala* was built by Atlas Aircraft Corporation initially a private sector enterprise financed by the South African government that soon became an Armscor subsidiary, ultimately evolving into subsidiaries of today's Denel. See Landgren, *Embargo Disimplemented*, 65-69, Helmoed-Römer Heitman, *Unpublished Manuscript* provided to the author in June 2003, 132; and http://www.denel.co.za, (12 March 2004).

60. Batchelor and Willett, *Disarmament and Defence*, 30. Atlas was transferred from Armscor to Denel in 1992 and was later renamed Denel Aviation. Still part of Denel in 2004, it was split into different subsidiaries of Denel's Aerospace Systems Group. See Denel *Annual Report 2004*, http://www.denel.co.za, (24 March 2005).

61. In the 1960s, the Defence Ordnance Workshop was recast as Lyttleton Engineering – *Lyttleton Ingeneurswese Werke* (LIW). Initially focused on small arms, it evolved substantially throughout the *apartheid* era and is now the subsidiary of Denel that produces South Africa's most sophisticated field artillery systems. Landgren, *Embargo Disimplemented*, 37-39, Heitman, *Unpublished Manuscript*, 132; Denel *Annual Report, 2004*, http://www.denel.co.za, (21 March 2005).

62. Landgren, *Embargo Disimplemented*, 38-9.

63. According to Landgren, prior to 1963 South Africa "spent 70 percent of its military budget on arms procurement overseas." Landgren, *Embargo Disimplemented*, 9.

64. Ibid., 41. In 1961 alone, South Africa obtained 127 licensing agreements for local manufacture of military items, a program that grew in subsequent years.

65. Ibid; Gavin Cawthra, "South Africa," in *Arms Procurement Decision Making, Volume II: Chile, Greece, Malaysia, Poland, South Africa and Taiwan*, ed. Ravinder Pal Singh (Oxford: Oxford University Press, 2000), 145.

66. Heitman, *Unpublished Manuscript*, 133. The two subordinate companies were PMP (Pretoria Metal Pressing, responsible for manufacturing small arms ammunition) and LIW (*Lyttleton Ingeneurswese Werke*, responsible for design and production of weapons systems, initially concentrating on small arms and smaller crew-served weapons).

67. An Act of Parliament provided the legal basis for this organization: The Armaments Development and Production Act, 1968 (Act No. 57 of 1968). This would remain the basic enabling legislation for Armscor until April 2004.

68. Although Armscor in 1992 transferred its production role to another public sector corporation created at that time (Denel), Armscor has survived to the present day and still is part of the South African defense establishment. It still is responsible for defense acquisition in South Africa. See, *inter alia*, Batchelor and Willett, *Disarmament and Defence*, 29-36, 126; Landgren, *Embargo Disimplemented*, 45-59; Armscor *Annual Report*, 2003-2004, and http://www.armscor.co.za, (25 March 2005).

69. Batchelor and Willett, *Disarmament and Defence*, 29-36, 126; Landgren, *Embargo Disimplemented*, 45-59; and Armscor *Annual Report*, 2003-2004, and http://www.armscor.co.za, (25 March 2005).

70. The South African Parliament had created the Council, a public sector body separate from the civil service, intended to perform pure and applied research, coordinate other national scientific research efforts and encourage the training of researchers and technicians. See Denys Graham Kingwill, *The CSIR-the first 40 Years* (Pretoria: Scientia Printers, CSIR, 1990), 8-25; Landgren, *Embargo Disimplemented*, 38-39. See also "South African Defense Industry," Global Security Organization http://www.globalsecurity.org/military/world/rsa/industry.htm, (30 June 2003).

71. The early mine protected vehicles included the *Casspir* and the *Mamba*, still available from the South African enterprise BAE Land Systems OMC, see "Mechem," *Journal of Mine Action*, June 2000, http://www.maic.jmu.edu/journal/4.2/Profiles/mechem.htm, (21 September 2004), and http://www.baesystemsomc.co.za, (5 April 2005).

72. McWilliams, *Armscor*, 11.

73. The French partners were Matra and Thomson CFS. The air defense missile was called the *Cactus* in South Africa, and first deployed operationally in South Africa in 1968. South African engineers were heavily involved in the design of the missile, but it is not clear that the *Crotale/Cactus* was ever produced in South Africa, although it has been a staple of French arms exports. This missile was one of the first effective short-range SAMs and still is widely used. See Landgren, *Embargo Disimplemented*, 106-108 and McWilliams, *Armscor*, 18.

74. South African Air Force officers told the author in 2003 that the incentive for the helmet sight was the fact that the South Africans could not build fighter aircraft sufficiently superior in performance to those of their Soviet-backed opponents to allow South African pilots to maneuver the aircraft into position for a traditional engagement. The helmet sight allowed the pilot to guide the missile to the enemy even when the aircraft was not in an optimum engagement position.

75. Oates interview, 12 June 2003.

76. Landgren, *Embargo Disimplemented*, 42; for contextual detail, see also Batchelor and Willett, *Disarmament and Defence*, 32.

77. Holloway interview, 21 June 2004.

78. Http://www.csir.co.za, (13 February 2003); http://www.armscor.co.za, (13 February 2003).

79. Landgren, *Embargo Disimplemented*, 42-4.

80. Ibid., 45-55; Batchelor and Willett, *Disarmament and Defence*, 32-40.

81. See, *inter alia*, Seegers, "South Africa's national security management system, 1972-1990, 253-273; and R. Rotberg, "Decision making and the military in South Africa," in *South Africa and Its Neighbors: Regional Security and Self Interest*, ed. R. Rotberg *et al* (Lexington, MA: Lexington Books, 1985), 13-26. Batchelor and Willett, *Disarmament and Defence*, 32.

82. Cawthra, "South Africa," 145. See also R. Matthews, "The development of the South African military-industrial complex," *Defence Analysis* 4, no. 1 (1988):7-24; Batchelor and Willett, *Disarmament and Defence*, 32.

83. Ibid. See also M. Brzoska, "Arming South Africa in the shadow of the UN arms embargo," *Defence Analysis* 7, no. 1 (1999):21-38; and Landgren, *Embargo Disimplemented*, 191-247.

84. Steyn et al, *Armament and Disarmament*, 73; *White Paper on South African Defence Related Industries*, 1999, Chapter 2, Paragraph 2 http://www.info.gov.za/documents/whitepapers/index.htm#1999, (24 March 2005).

85. *White Paper on South African Defence Related Industries*, 1999, Chapter 2, Paragraph 2 http://www.info.gov.za/documents/whitepapers/index.htm#1999, (24 March 2005).

86. Batchelor and Willett, *Disarmament and Defence*, 37-9.

87. See, *inter alia*, T. Shaw and E. Leppan, "South Africa: white power and the regional military-industrial complex," in *Southern Africa in the 1980s*, ed. O. Aluko and T. Shaw (London: George Allen & Unwin, 1985), 251-77.

88. Landgren, *Embargo Disimplemented*, 18.

89. The frigates were designated the *President* class in South Africa. They were acquired in 1963 and 1964. The submarines were French *Daphne* class, purchased in 1967 and actually acquired in 1970 and 1971. Heitman, *South African War Machine*, 80-1.

90. Landgren, *Embargo Disimplemented*, 70-4.

91. Ibid., 68-70. The jet trainer was the Italian-designed Aermacchi MB-326, locally called the *Impala*, employed by the South Africans in a ground attack role. The aircraft was equipped with a Rolls-Royce *Viper* engine, a collaborative UK-Italian design. Atlas also produced light propeller aircraft based on modifications of foreign designs. These included the *Bosbok* (based on the Aeritalia AM-3C) and the *Kudu* (based on the Aermacchi-Lockheed AL-60).

92. Ibid., 75-7. The South African role in the manufacture of these aircraft at this early date still is somewhat of a mystery. Atlas Aircraft Corporation (with associated subcontractors) ultimately was able to manufacture most of the needed replacement parts. By 1985, Atlas was testing the prototype of an indigenously designed gunship based on the French-designed *Alouette* III and was producing the *Oryx* medium-lift utility helicopter, a South African version of the French-designed *Puma*. By the early 1990s, the South Africans had designed a new attack helicopter derived in large part from the *Puma*.

93. The *Ratel* was based on a German design but significantly modified and equipped with a powerful and robust imported bus engine (to avoid sanctions). According to Landgren, the new combat vehicle was inspired by a prototype from the West German Bussing company (later amalgamated with Maschinen Fabrik Augsburg-Nurnberg-MAN). In South Africa, the *Ratel* was designed and produced by the private sector company Sandock-Austral, following the precedent of the earlier *Eland* contract. Landgren, *Embargo Disimplemented*, 24; Savides interview, 18 June 2003.

94. The truck family was called the Samil ("South African Military") series. Much of the original technology was borrowed from European manufacturers of civilian equipment such as Unimog and Magirus Deutz, although the mine-defeating technology had its origins in the Rhodesian War. Heitman, *Unpublished Manuscript*, 132-3. See also Landgren, *Embargo Disimplemeted*, 81-2, 96-7 and McWilliams, *Armscor*, 31, 39-40.

95. McWilliams, *Armscor*, 82-3.

96. Ibid., 55. The South African versions were designated the R-4 and R-5. The *Galil* itself is a refinement of the Soviet Kalashnikov assault rifle. Its main advantage was its chambering in the new NATO standard 5.56mm ammunition.

97. The South Africans called their *Centurion*-derived main battle tank the *Olifant* ("elephant"). Landgren, *Embargo Disimplemented*, 39, 85-7.

98. The *Rooikat* was unveiled in 1988 and in series production by 1989. McWilliams, *Armscor*, 54.

99. This was the *Valkyrie* 127mm multiple rocket launcher mounted on a Unimog chassis, apparently inspired by the Soviet-designed BM-21. Series production of this weapon commenced in 1980. Landgren, *Embargo Disimplemented*, 88.

100. The G-5 is a towed version of the gun, the G-6 is self-propelled. They were built by the Armscor subsidiary Lyttelton Engineering (LIW). With their associated South-African designed base-bleed munitions, they were (and still are) among the most effective extended range field artillery systems in the world. For additional details on the origin of the G-5 and G-6 guns, see Chapter 4. The basic gun system actually was the invention of the Canadian-American Gerald Bull, though the South African version was significantly refined. The wheeled, self-propelled chassis for the G-6 is an innovative indigenous design. The story of Gerald Bull's (and his Space Research Corporation's) participation in the technology transfer of the gun system is complex and somewhat tragic. For details, see William Lowther, *Arms and the Man: Dr. Gerald Bull, Iraq and the Supergun*, (Novato, CA: Presidio Press, 1991), 113-42; and McWilliams, *Armscor*, 71-2.

101. Heitman, *War in Angola*, 40-2, 49-50, 138-9, 143-4, 344; McWilliams, *Armscor*, 73, Bridgland, *The War for Africa*, 87.

102. The *Oryx* combines a standard *Puma* fuselage with a "Super *Puma*" engine and tail boom, along with an advanced South African avionics suite, giving it impressive capabilities. It was originally built in South Africa using *Puma* airframes acquired from Romania, and then equipped with South African avionics and other components. Foreign pilots who test fly the *Oryx* typically give it very high marks for high performance and user-friendly operation. Interviews with Major General Winston Thackwray (South African Air Force, retired) Advisor to the Managing Director EADS South Africa, 21 June 2004; Heitman interview, 23 June 2004 and personal communication with the author, 20 April 2005.

103. The *Cheetah*, an upgrade of the *Mirage* 3 performed by Atlas Aircraft Corporation, bore an uncanny resemblance to the Israeli *Kfir*. The C variant apparently was an Israeli-built *Kfir*. Landgren, *Embargo Disimplemented*, 63, 65-6; Heitman, personal communication with the author, 20 April 2005.

104. These had been acquired in 1963 and 1964 and the modernization took place between 1968 and 1973. Heitman, *South African War Machine*, 80-1.

105. This was the *Drakensburg*, designed and built in South Africa. McWilliams, *Armscor*, 44-5; Landgren, *Embargo Disimplemented*, 119.

106. Ships purchased from Israel were the *Reshef*-class missile boats and *Dabur* patrol craft, the former later produced in South Africa as "*Minister*-class." The South Africans also acquired the Israel *Gabriel* anti-ship missile, which subsequently was produced in South Africa as the "*Skerpioen*" (Scorpion). McWilliams, *Armscor*, 54-5; Landgren, *Embargo Disimplemented*, 218-22. See also Heitman, *The South African War Machine*, 84-5.

107. See Landgren, *Embargo Disimplemented*, 153, 164-5. South African military personnel that served in the Defence Force in the *apartheid* years repeatedly have mentioned the Israeli connections to the author, though they are reluctant to divulge details.

108. The air-to-air missile program began in 1966 and resulted in the V3B *Kukri* first employed successfully in 1972, a short-range infrared homing missile apparently based on the Matra R-550 Magic, but slaved to the pilot's helmet sight. Helmoed-Römer Heitman, *War in Angola*, 355-6; and Heitman, *South African War Machine*, 128, 132. The ground-to-air missile was the *Cactus/Crotale* described earlier in the chapter. The anti-armor missile was a wire-guided system similar to the US TOW. The ship-borne anti-ship missile was a derivation of the Israel *Gabriel* missile described earlier, locally called the *Skerpioen*.

109. Oates interview, 18 June 2003. For historical detail, see also Bridgland, *The War for Africa*, 150-60.

110. The guided bomb was designated the H-2. Steyn et al, *Armament and Disarmament*, 74; Heitman, personal communication with the author, 12 June 2004.

111. McWilliams, *Armscor*, 44-6, 82-3.

112. See, *inter alia*, McWilliams, *Armscor*, 44-6; Heitman, *War in Angola*, 356 and personal communication with the author, 20 April 2005. Heitman says these munitions were an unusual assembly of ball bearings in a composite matrix and cased in plastic, custom tailored to the intended target. This application seems to have provided much more uniform projectile dispersal and more predictable results than aerial bombs manufactured elsewhere at the time.

113. These were Somchem and Naschem, currently divisions of Denel Ordnance. See Batchelor and Willett, *Disarmament and Defence*, 30; and http://www.denel.co.za, (3 April 2005).

114. Steyn et al, *Armament and Disarmament*, 76. For some additional detail, see Chapter 4.

115. Interview with Paul J. Gerber, Group Manager for Business Coordination and Counter-trade, Grintek Group Limited, 12 June 2003.

116. See Heitman, *War in Angola*, 355.

117. See Chapter 4. The South Africans flirted briefly with the prospect of an artillery-delivered nuclear warhead, although it appears that this program did not get beyond the conceptual stages. http://nuclearweaponarchive.org/Safrica/SABeginning.html. (22 March 2005). For details of the South African chemical and biological weapons development program, see Chandré Gould and Peter I. Folb, "The South African Chemical and Biological Warfare Program: An Overview," *The Nonproliferation Review* 7, no. 3 (Fall-Winter 2000): 10-23 and Stephen Burgess and Helen Purkitt, *The Rollback of the South Africa's Chemical and Biological Warfare Program* (Maxwell AFB, AL: USAF Counterproliferation Center, Air War College, 2001). For details on South Africa's nuclear weapons program, see Helen E. Purkitt and Stephen F. Burgess, *South Africa's Weapons of Mass Destruction* (Bloomington, IN: Indiana University Press, 2005); Steyn et al, *Armament and Disarmament*; Roy E. Horton III, *Out of (South) Africa: Pretoria's Nuclear Weapons Experience* (Colorado Springs, CO: USAFA Institute for National Security Studies, Occasional Paper Number 27, August 1999).

118. One source that requested anonymity described to the author a protracted discussion in South African military circles in the 1980s of an air-delivered nuclear strike against Cuba involving in-flight refueling of the strike aircraft, though it is unclear if this idea progressed beyond speculation about possibilities. Heitman says that the South Africans configured five Boeing 707s for these new roles. Three were designed for both Electronic Intelligence (ELINT) collection and refueling, one was a single-purpose platform and one was used for training. The Israelis apparently performed the tanker (refueling) conversions, while the South Africans seem to have done most, if not all, of the ELINT manufacture and installation. Heitman notes that the South Africans had worked on a side-looking airborne radar (SLAR) capability for these aircraft. Heitman, personal communication, 20 April 2005.

119. Heitman, *Unpublished Manuscript*, 133. The space program was coordinated by (the Armscor subsidiary) Houwteq using Armscor's Overberg Test Range near Bredasdorp in the Western Cape. State President F.W. de Klerk dismantled the nuclear program in 1990 and space programs in 1993. See also Batchelor and Willett, *Disarmament*

and Defence, 95; and Steyn et al, *Armament and Disarmament*, 95-103.

120. Heitman, *Unpublished Manuscript*, 65.

121. Information provided by Johan van Rensburg, US Commercial Service, personal communication with the author, 12 June 2004. Rensburg characterizes the industry of the late *apartheid* era as "wide but shallow," able to accommodate a variety of different requirements but not to the depth or sophistication of arms industries in developed countries.

122. Heitman, *Unpublished Draft*, 65.

123. *White Paper on South African Defence Related Industries*, 1999, Chapter 2, Paragraph 15, http://www.info.gov.za/documents/whitepapers/index.htm#1999, (24 March 2005).

124. Ibid. Paragraph 17.

125. Batchelor and Willett, *Disarmament and Defence*, 81.

126. Ibid., 33.

127. Landgren, *Embargo Disimplemented*, 175-87.

128. UNSCR 558, rescinded on 25 May 1994.

129. Landgren, *Embargo Disimplemented*, 173-87.

130. Batchelor and Willett cite studies arguing that heavy investment in arms industries was "misinvestment," reflecting choices based on strategic considerations rather than market factors, Batchelor and Willett, *Disarmament and Defence*, 33, 42. See also Cawthra, "South Africa," 145. Landgren says that South African defence spending rose about 30 percent per year between 1970 and 1975 and about 20 percent from 1976 to 1980. She says that "arms production per se absorb[ed] the largest part of the defence budget." Landgren, *Embargo Disimplemented*, 46-8.

131. Batchelor and Willett, *Disarmament and Defence*, 46-8. See also David Fig, "Apartheid's Nuclear Arsenal: Deviation from Development," in *From Defence to Development: Redirecting Military Resources in South Africa*, ed. J. Cock and P. Mckenzie (Cape Town: David Phillip, 1998), 163-180.

132. Ibid., 62-3. See also S. McMillan, "Economic Growth and Military Spending in South Africa," *International Interactions* 18, no. 1 (1992): 35-50. Heitman cautions, however, that South Africa's defense budget in the 1970s and 1980s typically ranged at about 4 percent of GDP, a high proportion but not prohibitively so. He attributes much of the recession in the 1980s to a growing reluctance on the part of foreign investors to invest in a country that seemed on the verge of a transition to black majority rule. Heitman, personal communications, 12 June 2004 and 20 April 2005.

133. See, *inter alia*, T. Moll, "Did the apartheid economy fail?" *Journal of Southern African Studies* 17, no.2 (1991): 271-291; and T. Moll, "From booster to brake? Apartheid and economic growth in perspective," in *The Political Economy of South Africa*, ed. N. Nattrass and E. Ardington (London: Oxford University Press, Press, 1990), 73-87.

134. Heitman, personal communication with the author, 12 June 2004; and interview, 23 June 2004.

135. Since the 1970s, the governments of both John Vorster and P.W. Botha had made a series of efforts to achieve internal solutions to the country's political problems; but by the late 1980s, the social and economic strains of domestic unrest and interminable regional conflict had begun to tell on South Africa's white minority, which increasingly saw little prospect of a better future. However, without the attenuation of the East Bloc threat, the South African government would have had difficulty convincing the white population to risk fundamental political change. Batchelor and

Willett, *Disarmament and Defence*, 62-6. Heitman, personal communication, 12 June 2004 and 20 April 2005. See also Friedman, *The Long Journey*.

136. See, *inter alia*, Scott Thompson, "South Africa and the 1988 Agreements," in *Disengagment from Southwest Africa: The Prospects for Peace in Angola and Namibia*, ed. O. Kahn (New Brunswick, ME: Transaction Publishers, 1991), 117-30; and John Marcum, "Retrenchment and Recalculation: South Africa and the Anglo-Namibian Agreements," in *Disengagment from Southwest Africa: The Prospects for Peace in Angola and Namibia*, ed. O. Kahn (New Brunswick, ME: Transaction Publishers, 1991), 131-48.

137. These efforts were facilitated by other dramatic events occurring at the same time – the collapse of Soviet hegemony in Eastern Europe, decreasing Soviet interest in Africa and Soviet disinclination to continue substantive support to regional clients. For an overview of the African political context at the time, see Michael Clough and Jeffrey Herbst, *South Africa's Changing Regional Strategy: Beyond Destabilization* (New York: Council on Foreign Relations, 1989).

138. See, *inter alia*, H. Adam and K. Moodley, *South Africa's Negotiated Revolution* (Cape Town: David Philip, 1992); Steven Friedman, Editor, *The Long Journey: South Africa's Quest for a Negotiated Settlement* (Johannesburg: Ravan Press, 1993); David Ottaway, *Chained Together: Mandela, De Klerk and the Struggle to Remake South Africa* (New York: Random House, 1993). For a discussion of de Klerk's motivations from his own pen, see F.W. de Klerk, *The Last Trek – A New Beginning* (New York: St. Martin's Press, 1999), particularly pages 159-341.

139. A finalized national constitution was adopted in May 1996.

CHAPTER 2

Majority Rule and a Transformation of "Security"

A Time for Change

The advent of the de Klerk presidency in 1989 dramatically affected the South African military establishment and the country's armaments industry. One of de Klerk's key objectives was an end to the country's regional conflicts, an objective at least partly motivated by a desire to reduce the expenses those conflicts entailed. His new administration was eager to begin, and the state's defense budget was cut by over forty percent between 1989 and 1993.[1] Defense procurement fell by 60 percent in the same period.[2] Arms production as a percentage of South Africa's manufacturing output was reduced from a high of 6.6 percent in 1989 to 4.5 percent by 1994.[3] According to Cawthra, "[defence industry] employment fell from 150,000 in 1989 to just over 70,000 in 1993, while the share of defence R&D as a proportion of the country's total R&D fell from 48 percent to 18 percent."[4] Between 1990 and 1993, South Africa also discontinued its programs to develop nuclear, biological and chemical weapons and ballistic missiles.[5]

One of the more interesting changes during this period was the devolution of the public sector arms industry, specifically Armscor.[6] While it remained the acquisition arm of the Department of Defence, Armscor gave up its production role and lost its arms-manufacturing subsidiaries.[7] A new, state-owned enterprise, Denel, was incorporated in 1992 to manage the various production parastatals and some of the test facilities.[8] This new corporation was moved out of the Department of Defence and into the Department of Public Enterprises. Cawthra notes that over 15,000 Armscor employees also were transferred to Denel at the time[9] and says that "both Armscor and Denel made efforts to become more transparent, representative and accountable," even to the point of "publishing annual reports for the first time in 1993 and appointing new board members."[10] These were dramatic changes for what had been a highly secretive community. But the transition was accomplished by executive fiat rather than parliamentary mandate. The new arrangement still limped along on the enabling legislation of 1968, leaving many legal ambiguities that would not be resolved until new legislation came into force in 2004.[11]

The decline in the defense budget between 1989 and 1994 had a significant impact on private sector arms producers. According to Botha, the government had no industrial transition program and the industry found itself "restructured in an *ad hoc*, market driven and largely chaotic way." With some prominent exceptions, the loss of government contracts was not balanced by corresponding increases in arms exports. Diversification into civilian products also proved much more difficult than anticipated.[12] While some diversification did occur, few firms achieved it.[13] After a traumatic series of rationalizations, downsizing and enterprise failures, the industry by 1994 was significantly smaller and less capable than it had been in 1989.[14]

Still, not all the news for the arms industry in the early 1990s was unfavorable. The radical decline of the domestic market had motivated the industry to seek export opportunities. Batchelor and Willett report that the value of South African arms exports grew by an astonishing 160 percent (in real terms) between 1989 and 1993, rising in that period from a mere 0.3 percent of total exports to 1.1 percent.[15] To the surprise of many in South Africa and elsewhere, newly elected South African President Nelson Mandela publicly defended the country's arms industry and promoted its products on his overseas trips.[16] The new ANC civilian leadership of the Department of Defence echoed his enthusiasm.[17] Minister of Defence Joe Modise was "particularly keen to promote a flourishing arms industry" seeing in it "an essential attribute of a strong sovereign state."[18] Then in May 1994, the United Nations lifted the embargoes on supplying arms to South Africa and buying arms from it.[19] Following the rescission of the embargoes, Armscor almost immediately announced its intention to vastly expand South African arms exports.[20] With the advent of majority rule and an end of UN sanctions, South Africa's corporate leaders at first were confident that eager foreign buyers would line up to buy their combat-proven military equipment. Industry analysts at the time also believed somewhat naïvely that the opportunity to diversify into non-defense products would sustain industries formerly devoted exclusively to armaments.

Despite a limited basis for optimism, a sober assessment in 1994 probably would have concluded that the advent of majority rule did not bode well for the defense industry as a whole. True enough, the industry still employed some 70,000 South Africans and was the country's premier exporter of manufactured goods,[21] but European and North American arms producers dominated the large overseas markets for arms, and were determined to maintain their advantages. The Cold War had ended, eliminating a major motivation for arms acquisitions on the world scene. In any event, breaking into the lucrative national markets of NATO countries, with their entrenched local industries and powerful advocacy groups, would have been a daunting prospect for any outside actor regardless of the quality of the product. Then, potential customers did not have the confidence that the South Africans could deliver the large-scale or long-term service of the established European and American defense industries.

Oversees market access was not the only significant challenge. With the prominent exceptions of the President and his Minister of Defence, many officials in the emerging ANC government had deep reservations about anything connected to the old regime, particularly its security services and the infrastructure that once empowered them, and some of South Africa's new leaders had ideological or moral qualms about an international arms trade.[22] Too, citizens in the attentive public increasingly were aware of the price they had paid since the 1960s for their government's investment in "security." Many saw little reason to continue it: South Africans now expected a "peace dividend" of improved infrastructure, housing and other government services, not a continuing investment in military materiel.[23] So no informed observer in 1994 could have predicted with confidence that the South African defense industries – and South Africa's capacity to produce state-of-the-art military technology – would survive the country's political transformation.

The future of the country's arms industry was plunged in further doubt by scandal in September 1994. Six months after the advent of majority rule, the new Mandela government was rocked by an international uproar when the world press revealed covert South African arms deliveries to Yemen, a country with a deplorable human rights record in the throes of a civil war. In the resulting turmoil, Mandela felt constrained to appoint an investigating commission headed by an eminent South African jurist, Edwin Cameron. The Cameron Commission's first report in June 1995 excoriated Armscor for lax supervision of exports, suggesting a number of measures to improve the oversight of the nation's arms sales. A second report in March 1996 went much further. It recommended parliamentary review (and veto authority) over arms exports and questioned the propriety of South African involvement in the international arms trade altogether.[24]

The South African cabinet reacted to the initial findings of the Cameron Commission in 1995 by appointing a ministerial commission headed by Minister of Defence Modise to examine the issue of arms sales with a mandate to recommend ethical guidelines and corrective policy. The Modise Commission offered its findings in August 1995. One immediate result was the elimination of Armscor's prerogative to make autonomous policy decisions on arms exports. Another was the investment of that authority in a tiered structure headed by a National Conventional Arms Control Committee (NCACC), while still another was the appearance of clear ethical guidelines for arms transfers in policy documents published after 1995.[25] The furor over arms sales had abated by 1996, though continuing South African exports to questionable customers somewhat belied the articulated policy and kept the scandal simmering at a lower level.[26]

Still, the arms sales scandals cast a pall that threatened the long-term survival of the South African arms industry. Between 1961 and 1989, the industry had developed as a coherent government response to a unique set of geopolitical circumstances that clearly no longer existed. South Africa now faced no significant conventional military threat on its borders, nor did it appear one would materialize in the foreseeable future. After the transition to majority rule, the former government and its coercive agencies were thoroughly discredited, and most South Africans were eager to leave the past behind. The ambiguous and ambivalent signals from the emerging new ANC government at first offered no clear public sector commitment to an armaments industry, particularly one now seriously tarnished by scandal. It was very uncertain whether the attentive public would be sympathetic to that pillar of the *ancien régime*.

Yet, a decade later, the armaments industry was not only thriving, but had managed to embed itself in the long-term security vision of the South African state. South Africans, with a few prominent exceptions, no longer seemed to question its relevance to the future of their country. Not only had the industry assumed a critical and growing role in the nation's exports, but was playing an increasingly important role in other aspects of national industrial development as well. It had entered the international arms market in significant partnerships with foreign arms industries, and had become more dependent on exports than upon domestic sales. Armaments sold by major western firms around the world now increasingly included state-of-the-art technology developed and built in South Africa. How could all this possibly have occurred in one decade?

The answer lies in a transformation that fundamentally altered South African ideas and institutions. It was mediated by a remarkable series of national consultations – perhaps the most democratic process of defense decision-making in modern history. The consultations were not primarily about the arms industry, of course, but about the kind of security appropriate to the new South Africa and the kinds of institutions that would provide that security; but it was in this context that the defense industry was able to carve out a role that would assure it a viable future.

A New Kind of Security for a New South Africa

Prior to 1989, the leaders of South Africa's *apartheid* state had sporadically engaged their domestic opponents, seeking some mutually acceptable political accommodation. Then, increasingly during the 1980s, members of government, South African academia, civil society and the political opposition had communicated with each other in a variety of ways about a prospective democratic future for the country. However, these efforts did not bear much tangible fruit until the end of the Cold War and the advent of the de Klerk Administration in 1989. Shortly after coming to office, de Klerk's new government had begun substantive discussions with its political opponents – particularly the African National Congress (ANC) – in anticipation of the majority rule that finally arrived in April 1994.

South Africa's new willingness to negotiate a mutually acceptable future did not mask the profoundly different political visions of the competing parties and individual actors. The options were complicated by the differing agendas and the necessity for difficult choices in the face of competing demands for access and resources. These differing visions had many facets, but issues of security consistently took a center place, and the dialogue was complicated by substantially different understandings of the meaning of "security" – reflecting deep disagreement among South Africans over what, exactly, should be protected and the means that should be employed in the process. Regardless of their ideological inclinations, the citizens participating in the consultations after 1989 understood that they would be obliged to make a number of difficult choices about their security establishment and the country's armaments industry.

For some years prior to the advent of majority rule in 1994, ANC intellectuals had been debating the nature of the security establishment appropriate to a new South Africa.[27] Many of these individuals had deep roots in academe – either in South Africa or the United Kingdom. They were well acquainted with emerging new worldwide conceptualizations of "security," with the most advanced thinking about roles and missions of security forces and models of democratic civil-military relations. They now endeavored to apply "cutting edge" thinking from the world's best scholarship to South Africa's unique situation and circumstances.[28] Despite a diversity of views, this influential group of political "insiders" ultimately arrived at essential agreement about the kind of military needed by the new South Africa.[29] Much of this effort was reflected in the debates of the Military Research Group, a rather loose collection of academics and intellectuals that coalesced in the early 1990s. This organization became essentially the ANC "think tank" on security.[30]

As the perspectives of the Military Research Group gelled in the 1990s, one of the most prominent features of its thinking was a strong emphasis on a broad new conception of "security" – a conception similar to the one promulgated at the same time by the United Nations Development Program (UNDP).[31] This approach focused attention on the human needs of people rather than the sovereignty or prerogatives of states. It emphasized safety of individuals from chronic menaces like hunger, disease and political repression as well as the disruptions in the pattern of daily life caused by war and other forms of violence. Although South African scholars were inspired by "human security" concepts imported from elsewhere, they also contributed substantially to the ongoing worldwide debates.[32] A prominent feature of their new approach was its rejection of any vision of "security" that stressed the coercive application of state power.

Yet the new model of "security" envisioned by the ANC intellectuals did not reject a role for traditional agencies of the state – intelligence, police and military. Their model assumed a place for the structures and equipment already in place in South Africa, and emphasized a "primary function" for the defense forces – preservation of national sovereignty and territorial integrity – roles requiring the maintenance of a conventional military establishment much like the one already in existence. However, in their model, the new military forces were to be under very firm and accountable civilian control – and would not be the exclusive province of unaccountable "securocrats" as in the *apartheid* years. Their thinking was reflected in subsequent restructuring of both the executive and legislative branches of government.

By 1995, South Africa's Department of Defence had been reorganized to include a civilian Secretary of Defence under a (civilian) Minister and Deputy Minister. The Secretary would have several key roles: he would serve as the Ministry's accounting officer; he would be the principal advisor to the Minister of Defence on issues of defense policy; and he would supervise the "Secretariat" – the large civilian bureaucracy that now ran the department and oversaw "defence" in South Africa.[33] As a result of the reorganization, the Department now would consist of three elements: the Secretariat, the South African National Defence Force, and Armscor (which remained responsible for materiel acquisition). In Mill's words, this new arrangement, "marked a radical departure from the past when a civilian minister . . . had headed a ministry that was totally dominated by the military, had no civilian representation or involvement of any consequence, and was subject to virtually no effective system of financial or policy accountability."[34]

The restructuring in the Department of Defence occurred at the same time that the Parliament was undertaking efforts to consolidate legislative authority over defense policy. What emerged in the legislative branch of government was a separate committee in both houses, Senate and National Assembly along with a larger Joint Standing Committee on Defence with membership from both houses. The Joint Standing Committee now had significant constitutional prerogatives to oversee policy, funding and programs, and by the late 1990s had significantly expanded the parliamentary footprint in oversight of defense activities. (Later, when the Senate transformed into the Council of Provinces, the role of legislative oversight passed largely to the Portfolio Committee on Defence.)[35]

In the thinking about security evident in the emerging new strategic approach, the country's orientation primarily would be defensive. The new South Africa would no longer threaten its neighbors – in fact, its military typically was to be used in circumstances of regional cooperation. And despite its commitment to the "primary function," the Defence Force also would be tasked to accomplish a wide range of human security objectives, including disaster relief and peace support operations.[36] Not all of the members of the Military Research Group were unanimous on these issues, of course, and Cawthra observes that its ultimate willingness to retain a large, conventional force was partly motivated by a desire to placate the existing South African military establishment. But regardless of its motivation, the emphasis developed by the Military Research Group subsequently was embedded in the interim Constitution of 1993, the ANC Manifesto of 1994, the permanent Constitution of 1996 and follow-on official documents that more specifically defined the new defense establishment and its oversight by the executive and legislative branches of government.[37]

The ANC intellectuals provided a general vision for new security institutions, but three key actors were particularly responsible for guiding that vision to adoption in South Africa through a difficult and complex political transition: Minister of Defence Joe Modise, his deputy Ronnie Kasrils, and Chief of the Defence Force, Georg Meiring. Despite their vastly different backgrounds, these men worked as an effective team. With his status as former commander of the ANC's military arm, Modise had impeccable "liberation war" credentials, position and influence within the ruling ANC political party. Moreover, he had the personality and vigor to pursue an aggressive agenda. Kasrils, former chief of intelligence for the ANC, was uniquely sensitive to many of the concerns of the political actors and groups, particularly those in Parliament, and was able not only to translate their concerns into Department of Defence programs, but also to convey an assurance of departmental sympathy to their agendas and equities. Meiring, who commanded the Defence Force from 1993 to 1998, played an effective role in establishing a collaborative relationship between the hierarchy of the *apartheid* era Defence Force and the key leadership of the incoming ANC government. He promoted that communication through the vicissitudes and strains of the transition itself.[38] Under the leadership of Modise, Kasrils and Meiring, multiple political "voices" were heard on the issue of defense transformation, yet the end result was a substantial, conventional defense establishment.

By 2005, white South African military officers who had lived through this transition were prone to aver that they could always "relate to the MK chaps" (in other words, had a rapport with the former "enemy" combatants within the ANC's armed wing- *Mkhonto we Sizwe* – due to their common experience as soldiers). This may have been true to some extent, but it bears all the earmarks of a developing myth serving to rationalize the integration of formerly hostile military groups into one Defence Force. (The myth implicitly offers a new "us" versus "them" categorization in which civilian political officials, regardless of background, are assumed to be the "outsiders," deficient in soldierly virtues.) The myth overlooks the considerable suspicions – even paranoia – shared by all sides at the beginning of the negotiation process in 1989. Meiring deserves particular credit for his role in attenuating that suspicion.

Defining an Arms Industry for the New South Africa

The Modise-Kasrils-Meiring leadership also allowed the armaments industries to define a renewed partnership with the new government and new Department of Defence,[39] a relationship that seems to have crystallized quite early in the political transition.[40] The attitude of the new government was reflected in an October 1994 speech by Kasrils, who touted the importance of the industry to South Africa's export economy, its role in creating jobs and its responsibility for maintaining the country's strategic autonomy.[41] Kasrils' strong endorsement was all the more remarkable in view of the arms export scandal that had erupted a month prior to his speech. Departmental enthusiasm for a robust national arms industry was evident throughout Modise's tenure as Minister of Defence and continued under that of his successor, Mosiuoa G. P. Lekota, who assumed the role of Defence Minister in June 1999. Yet even though the new civilian leadership in the Department of Defence supported a strong national arms industry from the outset, the fundamental changes on issues of security in South Africa were mediated though genuine, wide-ranging consultation.

According to citizens who lived through these events, the effective transition of both defense and defense industry probably would not have worked without the active cooperation of two other key groups: South Africa's senior military leaders and the top managers of the arms industry. Neither group was completely agreed on political issues in the early 1990s. There were individuals in each that actively opposed political change, but both groups as a whole sought to reassure the emerging national leadership of a genuine willingness to be effective contributors to a new South Africa. This particularly was true of Defence Force Chief Georg Meiring, who commanded the Defence Force through the transitional period. Political and military insiders close to the events at the time seem generally agreed that he played a particularly important role in connecting the "old" Defence Force to the leadership of the "new" South Africa. His retirement in 1998 seems to have been the result of an obscure intrigue among ANC insiders, and his departure may have been orchestrated by his political foes. However, Meiring played an essential role in facilitating the military aspects of dramatic political change, and on the occasion of his retirement enjoyed South African President Mandela's sincere commendation for his honorable service.[42]

South Africa's captains of industry also played an essential role. The senior leaders of the arms industries – both parastals and private sector – seem to have been more willing to embrace the political changes than their military counterparts. In fact, "big business" in South Africa had since the recession of 1980s become increasingly vocal about the need for political reform.[43] Yet, during the actual transition period from 1989 to 1994, industry leaders remained somewhat in the background and were not prominent either in the ANC or its opposition. (There was no industry equivalent, for instance, to the oppositionist political roles played by former generals Hartzenberg and Viljoen.[44]) On the other hand, retired South African military leaders had achieved high positions in the defense industries and now are recalled as influential in mediating the industry's new status. Industry insiders give particular credit to retired Air Force Major General Julius Kriel, who rose in this period to the leadership of the key arms industry advocacy group in South Africa.[45] Kriel is said to have played a decisive role in persuading arms industry actors to collaborate and coordinate with each other and with the new government during the transition period, allowing the industry to speak with a unified voice in the consultative process.[46]

The new South African government that came to power in 1994 was committed to stakeholder "buy-in" on policy issues and sponsored a series of conferences that defined both the military establishment and the associations between the government and the defense-related industries. These consultations provided industry leaders an opportunity to define and "sell" the new affiliation, an opportunity that they quickly seized, though its success required vision and perseverance. But more importantly, the emerging new partnership also required a willingness of industry leaders to adapt to the priorities of the new South Africa. This they did with surprising alacrity. In 2004 industry leaders and public sector policymakers considered that this relationship still remained in a process of transition,[47] but the national consultations between 1996 and 1999 clearly had developed a consensus that now was firmly embedded in the country's political culture.

The consultations also ultimately led to three policy documents that defined and documented the new connections. The documents themselves are important, but because of the "buy-in" achieved amongst various stakeholders, the consultation probably was more important than the documents themselves. Still, each document made a notable contribution and is worthy of brief review. These are the *South African White Paper on Defence* (1996), the *South African Defence Review* (1998) and the *White Paper on South African Defence Related Industries* (1999). The documents provide both a roadmap for national policy and an interesting glimpse of the emerging collaboration between policymakers, industrial leaders and the attentive public in the new South Africa over a five year period. Each also has continuing significance for the South African armaments industry.

The White Paper on Defence

Because of the massive redirection of funding initiated by the de Klerk government after 1989, the military services had faced some years of austerity by 1994. That and the international sanctions of the previous decades had rendered much of South Africa's military materiel depleted, deteriorated and obsolete. By 1994, the military establishment was still there, but each of the services was facing severe problems of obsolescent materiel and declining capital investment.[48] One of the early challenges faced by the new government was a plea from the military for substantial recapitalization. Soon after the advent of the new government, the leaders of the military establishment, both civilian and military, began lobbying for a fundamental re-equipping of the armed forces.[49] They would be obliged to wait until 1998 before the government would commit itself to a major program of materiel acquisition, but their pressure contributed to a substantial South African effort (begun by 1995) to redefine the national military and its supporting armaments industry.

Parliament responded to the plea by demanding that the Defence Force first define itself and its role, demonstrating how it would embody the priorities of the new South Africa, and in response, in 1995 the Department of Defence launched an extraordinary, wide-ranging series of consultations with scholars, practitioners, civil servants and members of the attentive public.[50] The result of the consultations was a "white paper" defining the role of the military in the country. A new *White Paper on Defence* finally appeared in 1996 in the wake of an excruciatingly exhaustive consultative process.[51]

During the consultations, several different agencies (including the military establishment itself) had offered competing versions, but it was the vision of the ANC intellectuals in the Military Research Group that ultimately prevailed. The principal drafter of the successful paper was University of Cape Town scholar Laurie Nathan. The new official white paper was remarkably similar to ideas he had published earlier.[52] Even so, Nathan's first iteration of this paper proved to be highly contentious and seems to have provoked accusations from Defence Force officers that it was "anti-military" in tone and substance. Ultimately, the senior leadership of the Ministry of Defence intervened to provide "guidance." Nathan was obliged to incorporate a large number of changes into the final draft, and the new paper reflected input from a broad range of sources. Parliament's Joint Standing Committee on Defence scrutinized early drafts of the paper and required Nathan to defend his work in three special sessions before the paper finally was accepted by the Cabinet and approved by Parliament in May 1996.[53] Despite the wide collaboration, the new white paper bore the clear imprint of the Military Research Group.

There was more than a little irony in Nathan's authorship. His earlier prominence as an anti-war activist would not have seemed a propitious qualification to define a new security establishment with a powerful conventional military. But he and his colleagues succeeded in crafting a document that called for a substantial national Defence Force, unambiguously subordinating the new military establishment to strong civilian control with substantial accountability to the nation.

The new *White Paper* was a clear articulation of a new philosophy of national defense. Its authors tried very hard to capture the essence of fundamental transformation they believed was required in South Africa. They started by redefining national security, characterizing it in the broadest possible terms.[54] They went on to argue that that security would be "sought primarily through efforts to meet the political, economic, social and cultural rights and needs" of the citizenry, implying a much more limited role for the security organs of the state than was formerly the case.[55] Yet despite its apparent embrace of a human security paradigm, the *White Paper* unambiguously endorsed a conventional military establishment, stating that the new South African National Defence Force (SANDF) would be a "balanced, modern, affordable and technologically advanced military force" and that its "primary role [would] be to defend South Africa against external military aggression."[56] At the same time, the *White Paper* was very careful to emphasize that the new military would have a "primarily defensive orientation and posture," and that South Africa's foreign policy would emphasize "regional security" and "military co-operation with southern African states in particular."[57] The document placed considerable emphasis on strong civilian control of the new military, with specified linkages of accountability and oversight that extended to the executive and legislative branches of government and even to civil society.[58]

The *White Paper* went far in addressing core interests and concerns of the new ANC-dominated government, and provided a philosophical framework for the nature of the new South African military. It thus satisfied the parliamentary demand that had precipitated it. However, it left open a number of questions about the size, structure and equipping of the force. It did not entirely reconcile its vision of a conventional, high technology military with its assessment of the most likely roles for the new force. And it did not address key issues relevant to security-related research and development, production and acquisition, issues that pertained very specifically to the defense industry in South Africa, still reeling from the effects of the political transition in the early 1990s, and the arms export scandal of 1994 (along with the subsequent harsh indictment of the Cameron Commission). These issues would be taken up by subsequent documents.

The Defence Review

With the initial *White Paper* in hand by 1996, the Department of Defence could turn to the details of the new military establishment. South Africa's military leaders were motivated by a pressing desire to define the future Defence Force, since a host of issues (including armaments acquisition) had been left in abeyance pending government agreement on the details of a new structure.[59] Again, the Department of Defence initiated an extraordinary process of national consultation, resulting in the publication of a *Defence Review* in late 1998.[60] This consultation ". . . featured three National Consultative Conferences, a host of Regional Workshops in all provinces, Public Hearings in Parliament. . . . It drew together a wide range of interest groups, including: academics, clergy, *industrialists*, media, *pacifists*, Parliamentarians, and members of the defence establishment." (emphasis added).[61] The Minister of Defence at the time claimed it to be the "most inclusive and transparent" process of its kind ever undertaken, a boast difficult to refute despite its somewhat self-serving tone. Key figures in drafting this paper were Air Force Major General Len Le Roux and Army Colonel (Dr.) Rocky Williams, both at the time active members of the South African National Defence Force. Both considered that their role was to comply with the general guidance of the Interim Constitution (of 1993) and recently published *White Paper on Defence*, reconcile the competing visions of interested political actors and achieve consensus among a diverse community of stakeholders.

The Le Roux/Williams partnership had all the elements of a novel. Le Roux, the consummate professional staff officer from the old South African Defence Force, phlegmatic and intellectual, worked with a core staff of senior officers from the four military services. Williams was an effusive and colorful character, who had served as a clandestine ANC agent in the *apartheid* era South African military before leaving to study for a doctorate in the United Kingdom. Williams now argued radical options against the more conventional approaches emphasized by Le Roux, the career staff officer. Both sought to reconcile the even more radical views of participants like Laurie Nathan, and both strove to make the ultimate document acceptable to a very diverse constituency, a requirement that demanded considerable political acumen.[62]

One of the most important relationships in the compilation of the *Defence Review* was the extensive involvement of the Parliamentary Joint Standing Committee on Defence, which insisted on exercising its oversight responsibilities, discussing the *Review* issues in great detail, while at the same time "learning the ropes" about military affairs. The relationships between parliamentarians and senior military leaders were at times strained by severe mutual suspicions, but the outcome was beneficial – both the legislators and their military counterparts learned much in a short period, mutual suspicions were at least somewhat attenuated and Parliament achieved a remarkable degree of "ownership" over the resulting agreements.[63]

The *Defence Review* built upon the foundational *White Paper*, specifying in some detail the nature and role of the new South African National Defence Force (SANDF). It dealt with force design, doctrine and deployment and defined issues of resources and accountability. However, for the purposes of this book, the most interesting part of the *Defence Review* dealt with acquisition management. The *Review* defined the South African defense industry and specified how it would work in tandem with the South African government to maintain the country's technological capability.[64] Equally importantly, it noted conditions under which foreign industries could market military materiel in South Africa.[65] The *Review* stated that the country "should not strive for self-sufficiency in arms development, but only limited self-sufficiency in key areas."[66] However, the *Review* clearly implied that South Africa would retain a significant indigenous arms production capacity to meet its strategic requirements.[67]

The acquisition policy specified in the *Defence Review* bore some resemblance to the close ties that existed between government and industry in the era before majority rule. However, there were now at least four key differences: much more of the arms industry would consist of private sector enterprises than was the case in the past, when parastatals produced much of South Africa's military materiel exclusively for South African use. Producers of defense materiel now would be encouraged to diversify into non-military products, and the country would even more aggressively market its defense technology abroad. Significantly, foreign industries would play a key role in South Africa's arms purchases.

The *Defence Review* reflected a clear recognition that South Africa needed partnerships with external industries. In a significant shift for its acquisition policy, the country no longer would strive for full self-sufficiency in armaments: its military materiel requirements would be supplied from foreign sources as a matter of routine rather than grudging necessity. The foreign connections now would be overt: in the 1970s and 1980s, the *apartheid* state had engaged in a substantial amount of industrial espionage and covert arms acquisitions abroad, practices that were no longer needed. The new document articulated the value of foreign connections for technology transfer and access for South African industries to the world market, including substantive two-way relationships that would permit a foreign access to South Africa's defense technology, a novel feature for the South Africans. This, too, was a shift from the much more secretive South African practices of the past. Additionally, South Africa's new leaders saw in defense acquisitions an opportunity to compel foreign arms suppliers to invest in South Africa and provide the country's industries with access to the wider world's technology. The mechanism to promote many of these objectives was the counter-trade "offset" requirement, an issue discussed in greater detail in the next chapter.[68]

White Paper on the South African Defence Related Industries

After the approval of the *White Paper on Defence* in May 1996, the South Africans began to flesh out and define the general provisions of that document. As we have just seen, one result was the *Defense Review* supervised by the Department of Defence and approved by Parliament in April 1998, but still another was a comprehensive study of the defense industry itself. This was initiated in August 1996, leading to the publication in late 1999 of a *White Paper on the South African Defence Related Industries.* An important objective of this new official document was to define how the country would maintain its essential strategic capabilities. The paper was commissioned less than a year after the publication of the two Cameron Commission reports that had strongly censured Armscor's management of arms exports.[69] (In fact, the Commission had recommended this new *White Paper*, although probably with a view to constraining rather than empowering the arms industry.)

One immediate South African government action in the wake of the 1994 export scandals had been removal from Armscor of responsibility for oversight of arms exports (a role Armscor had played since the late 1970s) and investment of that control in several different bodies, the highest being the National Conventional Arms Control Committee (NCACC). This committee was made directly responsible to the Cabinet, and consisted of the ministers of defence, foreign affairs, trade and industry, safety and security, general services, the arts, culture, science and technology (along with the deputy ministers of defence, foreign affairs, intelligence services and safety and security).[70] The government clearly wanted to vet prospective arms sales very widely in an effort to avoid the embarrassing scandals of the mid 1990s. Senior Party officials outside and inside the defense community now would oversee arms exports as a matter of routine.

It was the NCACC rather than the Department of Defence that now was directed by the Cabinet to oversee the new *White Paper on South African Defence Related Industries*. This new document was compiled somewhat differently than its predecessors. It was not the result of the same kind of widely publicized, multi-sector consultative process; rather, its content reflected the perspectives of a more select working group with representatives from government, industry, academe and non-governmental organizations.[71] The Chief of Defence Acquisition, Shamin ("Chippy") Shaikh, chaired the working group's Steering Committee. In addition to the government and industry representation, the committee included academics Laurie Nathan (University of Cape Town) and Gavin Cawthra (University of Witwatersrand) scholars prominent in earlier consultations on Defense restructuring. Another scholar, Peter Batchelor, did much of the actual drafting of the paper.[72]

This new *White Paper* analyzed the South African arms industry, ranging across a wide spectrum of security and technical issues. It offered a comprehensive overview of the history and current status of the industry, a sector never previously exposed to such public scrutiny. The document also provided a very detailed historical analysis of the interests and involvements of the South African government in its arms industry. The analysis offered an intellectual rationale for continuing that relationship, with more specificity and substance than any of the preceding documents. But its most important feature was the opportunity it gave the armaments industry to collaborate with the government in defining its role in the new South Africa. The *White Paper* thus tied the arms industry and its leadership into a comprehensive national security vision.

In fact, this document is so fundamental that several of its features are worthy of particular note here. It reflected the South African government position "there is no clearly defined 'defence industry' [in South Africa],"[73] preferring the phrase "defense related industry."[74] (This terminology recognized the fact that producers of armaments in South Africa characteristically were embedded in industries that also offered non-military goods and services and included enterprises whose main products were other than arms.) The paper unambiguously declared that the government "has a duty to exercise control over any product or technology which can be termed an armament."[75] It was careful to claim that South Africa has an "industrial strategy" for its defense-related industry, but that this strategy applied only to "those *key technologies and capabilities* which [were] considered by the government to be strategic for national defence." [emphasis added].[76] It recognized defense related industries as an "integral part of South Africa's defence capability"[77] and expressed a willingness to support them with government resources, though with the *caveat* that the South African government "will be very selective of which technologies are to be retained."[78]

The *White Paper* looked internally at South Africa's security needs and externally at its export opportunities. It devoted considerable attention to the future roles and materiel needs of the military establishment, and it took up the issue of defense industry restructuring, including future privatization of Denel. It addressed various issues of labor relations and affirmative action. It also discussed the marketing and export of the nation's arms; delving into issues of non-proliferation and specifying the need for a Marketing Support Board to coordinate a coherent, comprehensive South African arms export strategy in which the government collaborated with public and private sector producers.

One additional item is of particular interest here. The document prescribed "strategies" for the country to safeguard its "essential defence technologies." These included government investment in private sector research and development, special projects and "advanced technology demonstrators."[79] It included the designation of "preferred suppliers" and technology "centers of excellence" within South Africa.[80] And significantly, it called attention to an entity called a Defence Evaluation and Research Institute (DERI). The concept already existed in the South African defense industry, but the new document gave it an institutional warrant. As envisioned in the *White Paper*, a DERI is small, high-tech group in close contact with its military consumers, yet distinct from the defense industry (though perhaps part of a larger commercial enterprise). It was to consist of a core of civilian scientists and technical experts that perform "leading-edge, high risk" research with little [immediate] commercial application. The DERI was to be funded by the Department of Defence and contribute particular expertise in narrow areas of military interest.[81]

For all of its significance, the new *White Paper* was much less prescriptive and much more descriptive than the earlier documents. It offered few recommendations at the end. Cawthra attributes this to the differing perspectives and agendas of the groups represented on the steering committee, which made agreement on specific, far-reaching recommendations problematic.[82] Still, it was a significant document that defined the past, present and (possible) future of the South African arms industry in a remarkable, multi-sector collaborative effort. Ironically, despite its novelty and importance, by 2004 industry insiders already were describing it as "dated" and in need of revision – an unintended tribute both to the rapid evolution of the arms industry after 1999 and the paper's success in embedding the concept of the government-industry partnership in the country's political culture.

Some Implications of the Documents

Taken together, the two *White Papers* and the *Defence Review* illustrate very comprehensive and careful efforts on the part of the South Africans between 1994 and 1999 to define their military establishment and provide for its material needs. These documents clearly indicate that the South African government now would serve as a fully engaged partner with industry in promoting arms exports, but exports were not portrayed as the major rationale for maintaining a significant arms industry. Rather, the documents assume the existence of continuing close ties between the arms producers and the national security establishment. They unambiguously announced South Africa's intention to maintain certain key security-related technologies for its own security needs, even if these required government subsidy. The military capabilities derived from a robust arms industry were deemed essential to South Africa's strategic autonomy. Despite the continuing vigor of private sector producers and the apparent intention to privatize some of the industry now found in the public sector (e.g. Denel) the South African government gave every indication that it intended to play a major role in all the activities of its defense-related industries, whether in maintenance of certain key technologies, purchase of its products, oversight of its research and development, or overseas marketing and sales of its products.

Still, the most significant implication of these documents, taken together, is the degree to which South Africa justified the maintenance of an arms industry by linking the arms producers to its own defense needs. This was captured in a locally popular phrase early in the debate: "no defence industry means no defence force."[83] The South African documents make no secret of the country's intention to pursue international arms sales, but at the same time, seem to recognize that arms exports still are a politically sensitive subject amongst the country's attentive public. Policymakers undoubtedly were concerned about the country's strategic options, but they also appeared to be using "national security" (or, more precisely, future South African strategic needs) as a rationale to deflect inevitable criticism of a robust national arms industry that exports much of its production.

This all leads to a possible discontinuity within the official documents. The anticipated range of future roles for the South African military establishment enshrined in the *White Paper on Defence*[84] and *Defence Review*[85] is somewhat at odds with the high-technology conventional military establishment also described in the *Defence Review*[86] and *White Paper on South African Defence-Related Industries*,[87] a military capable of defeating sophisticated external foes in classic military confrontations. The actual roles likely to be required of the Defence Force do not appear fully compatible with the high-technology arms that South African industries would be encouraged to develop in order to preserve South African strategic autonomy.

Since the publication of the *White Paper on Defence* and the *Defence Review*, the South Africans also have published a military strategy, which lists the missions envisioned for the Defence Force.[88] These missions cover a considerable spectrum, ranging from high to low intensity conflict. They include "high-end" roles such as "repelling [a] conventional onslaught" or "repelling [an] information onslaught" and such "low-end" missions as various kinds of peace operations, search and rescue and support of the national police.[89] While South Africa clearly seeks the capability to cope with all conceivable contingencies, including a massive, conventional attack by a sophisticated enemy on its borders, none of the documents really makes a case for the existence of a threat that would warrant such a capability. It is not at all clear that South Africa will face any such "high-end" threat in the foreseeable future.

In circumstances of plentiful resources, this disconnect of "ends" and "means" would not be a matter for concern, but South Africa's resources are far from infinite. Government funding needed to support a viable "high end" military capability may significantly attenuate the assets required for the more likely "low end" missions. Since the advent of majority rule in 1994, the typical employment of the Defence Force has been involvement in complex humanitarian emergencies, including a peace enforcement intervention (under a regional mandate) in Lesotho in 1998, flood rescue and relief in neighboring Mozambique in 2000, deployment of peacekeeping ground forces to Burundi under a bilateral arrangement, participation in the UN peacekeeping operation in the Democratic Republic of the Congo and several smaller-scale observer missions. By 2005 the expense of involvements in regional peace support operations had severely strained South Africa's financial resources.[90]

While the *White Paper on Defence* and a later *White Paper on South African Participation in International Peace Missions* (1999) specify conditions under which South Africa will engage in international peace operations,[91] none of the aforementioned documents explain how a high technology Defence Force organized primarily for conventional, high-intensity conflict along the country's borders will provide the wherewithal most appropriate to the articulated range of future roles, particularly against threats at the low end of the technological spectrum, including the challenges of various humanitarian operations.[92] In fact, this operational dilemma already had publicly surfaced by 2005. Analysts in the Defence Force were complaining about the country's lack of appropriate military materiel to support the peace operation in the ongoing UN mission in the Democratic Republic of the Congo.

The demands of low-intensity conflict and peace operations aside, Heitman argues that recent history has provided South African leaders an ample rationale for retaining a substantial conventional force. He says that the country's decisions about military capabilities in the mid 1990s were significantly colored by memories of the rapidity with which the East Bloc was able to deploy large numbers of conventional forces to African contingencies in the 1970s. For instance, deployments to Angola in 1975 grew quickly from one Cuban ground force division to two, and then ultimately to about 50,000 troops supported by sophisticated armor, firepower and air assets. Shortly after the intervention in Angola in 1975 and 1976, the Soviets also rapidly airlifted a multi-brigade Cuban force into Ethiopia, allowing the regime of Mengistu Haile Mariam to decisively repel a 1977 invasion from neighboring Somalia. Such rapid reinforcements of regional clients demonstrated to South African strategic planners that they could face a significant conventional threat in their region on relatively short notice given the mobility and lift of a hostile superpower. Too, recent African wars (like that in the Democratic Republic of the Congo from 1998 to 2002) have seen mid-intensity conflict, involving the use of armor, artillery, attack helicopters and high-performance combat aircraft. These wars also have included sophisticated riverine and lake operations. So it is entirely conceivable that even a peace enforcement operation in Africa could require substantial conventional ground, sea and air weaponry.[93] South Africa's unique military capabilities across the conflict spectrum are well known to its African neighbors and have resulted in regional pressures to contribute in substantial ways to local security initiatives.

The South African "defense" documents, though impressive results of hard work and tedious consultation, could not possibly have fully and permanently settled all the security dilemmas of the state. Their authors seemed to recognize them as authoritative for the requirements of the time, but also as snapshots of work-in-progress rather than definitive descriptions of South Africa's future. By

2004, industry insiders already were characterizing the *White Paper on Defence Related Industries* as somewhat obsolete – a backhanded tribute to the progress of government-industry partnership and the growing confidence of the private sector arms producers. By mid 2005, the South Africans were writing a new *White Paper on Defence* and *Defence Review*. Knowledgeable insiders suggested that one of the main revisions would be a significant restructuring of Armscor while another would shift the earlier emphasis on conventional military capabilities to those required for a greater capacity to respond to complex humanitarian emergencies.

Whether or not the South Africans actually will require substantial conventional capabilities to deal with future threats, they have identified essential "strategic capabilities" and state clearly in their official documents that they will seek to maintain them by direct government subsidy if necessary. Still, it is as yet unclear if they will have the resources and political will over the long term to follow through on their published intention. At present, the indications are ambiguous, for in fact, the issue of resources continued to be a troubling one as the 1990s gave way to the early years of the new century. Shortfalls continued to plague South African military planning, to the point that some key ongoing efforts to field newly purchased weapons were in jeopardy by 2005 and South Africa was having difficulty funding its regional military commitments along with routine training and maintenance.[94] The high cost of purchasing and fielding the new equipment in the "defence packages," actually may work against a deep official interest in futuristic military technologies. It appears increasingly unlikely that South Africa's leaders will be motivated to finance exotic domestic military technologies that provide little immediate economic or security return. Even so, the South African government clearly intends to control and exploit the country's arms production in partnership with its defense industry, so it is appropriate to note briefly the key mechanisms for doing so.

Government and Industry: Oversight and Partnership

Security policy in contemporary South Africa is a joint responsibility of the Executive and Legislative branches of government, a relatively conventional arrangement for a liberal democracy. Defense acquisition, the arms industry itself, and arms exports are all supervised by multiple agencies in the South African state. Significant responsibilities are delegated to ministries within the executive branch, again an unremarkable democratic norm. As Head of State and Commander in Chief of the armed forces, the President bears final responsibility for arms acquisitions, although the Minister of Defence has the "ultimate political authority and responsibility for the acquisition function."[95] More particularly, the President chairs a Cabinet committee concerned with defense and security policy. This Committee, or derivative components, makes collective decisions on military acquisition issues.[96] As discussed earlier in the chapter, an inter-ministerial National Conventional Arms Control Committee (NCACC) must approve arms exports.

South Africa's legislative branch of government also bears significant responsibility for oversight of defense-related issues, including arms industry production. Security issues are the primary concern of several permanent committees of Parliament, each of which has some bearing on defense acquisition. The most important early body was the Joint Standing Committee on Defence charged by the Constitution with responsibility to "investigate and make recommendations on the budget, functioning, organization, *armaments*, policy, moral and state of readiness of the SANDF." (emphasis added).[97] The Joint Standing Committee played a major role in shaping early documents such as the *Defence Review*. By 2005, parliamentary oversight of defense was exercised through a Select Committee on Safety and Security, responsible for oversight of defence, intelligence, police, justice and correctional services; and most importantly by a Portfolio Committee on Defence charged with monitoring and investigating, among other things, defence budgeting, rationalization, restructuring, functioning, organization, structure and personnel.[98] These mechanisms for security sector oversight are very much in consonance with the norms of liberal democracies.

It is within the executive branch itself that the South African organizational model displays at least one feature not typical of parliamentary democracies. The Minister of Defence oversees a department with three primary actors: a Defence Secretariat (essentially the Department's permanent staff), the South African National Defence Force (the military establishment itself) and Armscor. The Defence Secretariat has substantial responsibility for procurement, a fairly normal function for such a ministry in a parliamentary democracy,[99] but it is the presence of Armscor that departs most particularly from the traditional model. Armscor still plays an essential role in South Africa as acquisition agent both for the Department of Defence and for other security-related government entities – the police and prison service.[100] Its current mandated function is to meet the needs of the Department of Defence for materiel and for defense technology, research, development, analysis, test and evaluation.[101] In its own mission statement, Armscor describes its responsibility for maintaining "key [domestic] industries and technologies" and claims a role in marketing and promoting "the local defence-related industry."[102]

Armscor's function is fairly well detailed in policy documents, but less evident in practice. Various South African corporate actors inside and outside the public sector (including Armscor) still seem to be competing for roles in policymaking on issues of defense acquisition. Mid-level South African military officers expressed to the author in 2003 and 2004 their frustration with the relevance of arms purchases, as (presumably) mediated by Armscor. Defense industry insiders remain somewhat skeptical of Armscor's supervisory and marketing capabilities.[103] Still, such South African perspectives may be parochial, overly self-critical and a bit unfair. Heitman argues that some South African military personnel still are wedded to older security models, and fail to acknowledge the suitability of South Africa's new approaches and arms to its new strategic perspectives.[104] Too, given the relatively smooth function of the old system in the mid 1980s, the novelty of the defense rearrangements in the post-*apartheid* South Africa, the ongoing competition to define them, and the relatively short time since they were broached, it should not be surprising that the public and private sector linkages still seem somewhat undefined and incoherent to the South Africans participating in them.

As in the past, the defense industry is connected to the government at various levels, and the most important current feature of the relationship is the close partnership between the two. This relationship bears some similarity to the *apartheid* years, and South Africans once took the close connection for granted even if details were state secrets, but there are significant differences in the new affiliation. Both the government and the industry now openly acknowledge the relationship. And more importantly, the industry – including its private sector component – now contributes to a mutual agenda as a full partner.

A Marriage with Occasional Spats

The close relationship between government and arms industry is epitomized in two different (but related) organizations, the South African Aerospace, Maritime and Defence Industries Association (AMD) and the South African Defence Export Support Organization (SADESO). While AMD is a private sector entity with close linkages to the public sector, SADESO is a public sector entity with close connections to the private sector. Each of these organizations is a good illustration of the public-private sector collaboration on arms production in South Africa.

AMD is an industry advocacy group, created by the defense industry leadership in the waning years of the *apartheid* era, but its function is far more than simple advocacy. Virtually all of South Africa's significant defense-related corporate actors (in both the public and private sectors) now are affiliated with AMD. It exists to provide its members with research, marketing information, access, venues for industry coordination and resolution of disputes.[105] The organization is itself clearly in the private sector, but it serves a very public role as a partner with the government in consultations on defense-related issues. For instance, it was represented in the steering committee of the working group that compiled the *White Paper on South African Defence Related Industries* published in 1999, and is a major participant in SADESO (described below). AMD's most important role probably is that of connecting, facilitating, focusing and conserving the networks of industry insiders and civil servants that traditionally have supervised the country's arms industries.[106]

AMD's role is complemented by SADESO, an official instrument of the South African state chartered in August 2002 to develop and facilitate the country's arms export strategy.[107] In the words of one corporate insider, SADESO is a body specifically designed to pursue the interests of the "South African Military-Industrial Complex."[108] It is a formal partnership between the Department of Defence, Armscor and AMD, with representation from various other government departments. The Secretary for Defence chairs it.[109] His SADESO deputies include the CEO of Armscor and the Chairman of AMD.[110] At one level down, the Chief of Defence Acquisition chairs the organization's executive committee. There is little subtlety in South Africa's published intention to be a serious player in the international arms market or in the intensity of its drive to achieve government-industry cooperation in this effort.

While the degree of government-industry cooperation in South Africa is remarkable, it is by no means trouble-free. As might be expected, there are resentments in private industry springing from perceptions of unfairly preferential treatment – managers of the smaller enterprises have argued that government programs (notably the Industrial Participation regimes described in the next chapter) favor the large parastatals and larger industries and discriminate against smaller private sector actors, sometimes as a consequence of nefarious deals.[111]

A related issue is the politics of South Africans arms purchases that can trump the carefully constructed arms acquisition processes described in the official documents. This was evident in the late 2004 announcement of South Africa's intention to purchase European-produced A400M military transport aircraft to replace its fleet of aging C-130 transports. The A400M is a new generation military transport to be built in the future by a consortium of producers, of which Airbus and EADS are key members. South Africa presumably would buy between eight and fourteen of the aircraft for its Air Force when they finally are produced in 2009. (As this is written in mid 2005, the aircraft exists only on the drawing boards, lacking even a functional prototype.) South Africa has an undeniable need for more military airlift, particularly as it increases its commitment to regional humanitarian relief and peace support operations. Too, the acquisition of the A400M apparently will provide substantial opportunity for South African arms producers, who believe they will play an important role in the design, production and maintenance of the worldwide A400M fleet in the future. However, this arms acquisition decision is based on an uncertain delivery schedule and appears to have bypassed the Department of Defence procurement process altogether. In 2005 it was being managed by the Department of Trade and Industry, an unusual situation for a military acquisition program.[112]

The reputed deficiencies of the National Conventional Arms Control Committee (NCACC) are more serious causes of friction between government and industry. This ministerial-level organization was instituted in 1995 to assure the careful vetting of South African exports (in the wake of the scandals investigated by the Cameron Commission). Arms industry enterprises seeking to export materiel must now secure an NCACC license. Industry insiders complain that the decision-making process is far from transparent and is occasionally arbitrary. Worse yet, it is very slow. According to knowledgeable local authorities, this tardiness has preempted some significant export opportunities, even in cases where the NCACC ultimately approved the sales. Given the fast-moving nature of global commerce in the early 21st Century, it is inevitable that enterprising businessmen would chafe at impediments to sales opportunities. At the same time, it is easy to sympathize with a South African government determined to prevent the politically damaging arms export scandals of the recent past. On balance, however, the evidence suggests that the NCACC suffers from bureaucratic inertia that has on occasion compromised the country's economic interests. This aspect of the government-industry partnership appeared to be somewhat dysfunctional in 2005.[113]

One area in which government-industry relations may have been substantially more productive in the past than at present is in support of research and development. During the *apartheid* era, Armscor provided generous funding to basic and applied research throughout the arms industry. Production enterprises were encouraged to apply for special allocations to explore high risk/high promise new applications. This approach apparently played a significant role in encouraging creativity and innovation, prompting the arms industry to explore technology that ultimately offered quantum increases in military capability. Regrettably for the South African arms producers, and despite the enthusiastic endorsement of the *White Paper on the South African Defence Related Industries*, such funding did not survive the political transition to majority rule. Industry insiders familiar with the previous practice speak glowingly of its merits and regret its contemporary absence.[114] They believe that South Africa's future competitiveness in the tough international arms market requires more substantive government encouragement of research than currently is the case.

The current relationship between the South African government and the country's arms industry is illustrated by two interrelated programs that together offer a useful case study of the developing collaboration: the counter-trade "offsets" and the large-scale South African Defense "Packages," the massive arms purchases first announced in late 1998. In the case of the latter, the full ramifications have yet to be fully seen, and the acquisitions have generated controversy in the country, but the purchases touch most of the issues and actors of particular interest in this book. These programs also highlight one of the true novelties of the post-*apartheid* arms industry in South Africa – the extensive and growing connections between South Africa's arms producers and their foreign partners. The book now turns to these issues.

Notes

1. Peter Batchelor and Susan Willett, *Disarmament and Defence: Industrial Adjustment in South Africa* (Oxford: Oxford University Press, 1998), 67.

2. David Botha, *South Africa's Defence Industry, Charting a New Course?* (Pretoria: Institute for Strategic Studies, ISS Occasional Paper 78, 2003), 1.

3. Batchelor and Willett, *Disarmament and Defence*, 81.

4. Gavin Cawthra, "South Africa," in *Arms Procurement Decision Making, Volume II: Chile, Greece, Malaysia, Poland, South Africa and Taiwan*, ed. Ravinder Pal Singh (Oxford: Oxford University Press, 2000), 147. See also Batchelor and Willett, *Armament and Disarmament*, 70-1.

5. For details of South Africa's chemical and biological warfare programs and their dismantling, see Stephen Burgess and Helen Purkitt, *The Rollback of South Africa's Chemical and Biological Warfare Program* (Maxwell AFB, AL: USAF Counterproliferation Center, Air War College, 2001); for additional details on these and South Africa's nuclear weapons program, see Helen E. Purkitt and Stephen F. Burgess, *South Africa's Weapons of Mass Destruction* (Bloomington, IN: Indiana University Press, 2005); Roy E. Horton III, *Out of (South) Africa: Pretoria's Nuclear Weapons Experience* (Colorado Springs, CO: USAFA Institute for National Security Studies, Occasional Paper Number 27, August 1999). For "insider" descriptions of South Africa's nuclear program, its origins, growth, dismantling, and for the motivations, see Hannes Steyn, Richard van der Walt and Jan van Loggerenburg, *Armament and Disarmament: South Africa's Nuclear Weapons Experience* (Pretoria: Network Publishers, 2003); F.W. de Villiers, Roger Jardine and Mitchell Reiss, "Why South Africa Gave Up the Bomb," *Foreign Affairs* 72 (November-December 1993): 98-109; J. Shearer, "Denuclearisation of Africa: the South African dimension," *Disarmament* 16, no. 2 (1993): 171-185; and D. Howlett and J. Simpson, "Nuclearisation and denuclearisation in South Africa," *Survival* 35, no. 3 (1993): 154-173. See also F.W. de Klerk, *The Last Trek – A New Beginning* (New York: St. Martins Press, 1999), 273-4; and Batchelor and Willett, *Disarmament and Defence*, 71-2.

6. For a detailed description of this process, see Batchelor and Willett, *Disarmament and Defence*, 85-101.

7. Botha notes that part of the pressure to divest Armscor of its production capability derived from the fact that it was expressly forbidden by law to produce non-military goods for the civilian market. David Botha, *South Africa's Defence Industry*, 2.

8. *White Paper on South African Defence Related Industries*, Chapter 2, Paragraphs 25-7, http://www.info.gov.za/documents/whitepapers/index.htm#1999, (25 March 2005).

9. Cawthra, "South Africa," 147.

10. Ibid. Batchelor and Willett, however, call attention to the "remarkable degree of continuity within [Armscor's] executive and management structures," noting that the preponderance of senior management remained "white, male Afrikaners who had all had long careers in the domestic arms industry." Batchelor and Willett, *Disarmament and Defence*, 87.

11. The new legislation was the Armaments Corporation of South Africa, Limited Act, 2003 (Act 51 of 2003) actually signed into law by the State President in April 2004.

12. Author's interview with Arthur J.R. Oates, Director of Business Development, Denel, 12 June 2003. See also Diane Abrahams, *Defence conversion in South Africa: A faded ideal?* (Pretoria: Institute for Security Studies, ISS Occasional Paper 51, 2001), 2-3.

13. Botha argues that the only significant conversion of defense technology to civilian usage was in aircraft components and mine-protected vehicles. He also provides a useful overview of the reasons for widespread failure to diversify. Botha, *South Africa's Defence Industry*, 9. See also *White Paper on South African Defence Related Industries*,

Chapter 2, Paragraphs 53-6, http://www.info.gov.za/documents/whitepapers/index.htm#1999, (25 March 2005).

14. Botha, *South Africa's Defence Industry*, 2. However, Botha argues that by this point the South African arms industry had become much tougher and more competitive on the world market.

15. Batchelor and Willett, *Disarmament and Defence*, 80.

16. Ibid., 146.

17. Ronnie Kasrils, "The Future of South Africa's Defence Industry: the Government's Perspective," in *South Africa's Defence and Security in the 21st Century*, ed. W. Gutteridge (Brookfield, VT: Dartmouth Publishing Company, 1996), 117-26.

18. Batchelor and Willett, *Disarmament and Defence*, 128-9.

19. These were Security Council Resolutions 418 and 558 respectively.

20. Batchelor and Willett, *Disarmament and Defence*, 128.

21. James Herlihy, "South Africa's Arms Exporter Seeks International Respectability," *New York Times*, 26 August 1994, 9-13.

22. Batchelor and Willett, *Disarmament and Defence*, 113-20.

23. Abrahams, *Defence conversion in South Africa*, 1-2.

24. Batchelor and Willett, *Disarmament and Defence*, 146-50.

25. Ibid., 148. These documents included the *South African White Paper on Defence* of 1996 (see particularly Chapter 7, for additional detail), http://www.info.gov.za/ documents/whitepapers/index.htm#1996, (25 March 2005) and the *White Paper on South African Defence Related Industries* (a recommendation of the Cameron Commission). See also Richard A. Bitzinger, *Towards a Brave New Arms Industry*? (New York: Oxford University Press, IISS Adelphi Paper 356, 2003), 46-7. One example of South Africa's new approach was its refusal in 1999 to sell its new *Rooivalk* attack helicopter to Turkey due to concerns over the latter's human rights record. Given the strenuous efforts by South Africa to market this aircraft, and the very limited success up to that point, the decision to forego that sale must have been painful.

26. See *Human Rights Watch Report*, 12, no. 5 (A) (October 2000), Http://www.hrw.org/reports/2000/safrica/Sarfio00-05.htm), (25 March 2005) which noted South African arms sales between 1996 and 1998 to India, Columbia, Pakistan, Congo (Brazzaville) and Algeria. See also Stefaans Brummer, "SA Arms-Dealing Underworld," (Johannesburg) *Weekly Mail & Guardian*, 2-8 June 1995, 9; Petra Thornycroft, "SA's Poison Gas Secrets Sold to Libya," (Johannesburg) *Mail & Guardian*, 7 January 1997; "South Africa: An Irresponsible Arms Trader?" *Global Dialogue* 4 (2 August 1999) available at http://www.igd.org.za/pub/g-dialogue/ Special_feature/trader.html, (3 May 2005); and Batchelor and Willett, *Disarmament and Defence*, 149-50.

27. These included scholars like Nathan Laurie, Greg Mills, Gavin Cawthra and Rocky Williams. See, *inter alia*, "Ready to Govern: ANC Policy Guideline for a Democratic South Africa," adopted at the National Conference, May 1992, http://www.anc.org.za/ancdocs/ history/readyto.hmtl, (25 March 2005); L. Nathan, "Beyond arms and armed forces: a new approach to security," *South African Defence Review* 4, (1992): 12-21; and B. Fanaroff, "A trade unionist perspective on the future of the armaments industry in South Africa," *South African Defence Review* 7 (1992): 10-4.

28. The South African professional literature at the time reflected a rich ferment in thinking about new definitions of "security." See, *inter alia*, M. van Aardt, "Doing battle with security: a Southern African approach," *South African Journal of International Af-*

fairs 3, no. 2 (Summer 1996): 13-28; and Hussein Solomon, "From marginalised to dominant discourse: Reflections on the evolution of new security thinking" in *Caring Security in Africa*, ed. H. Solomon and M. van Aardt (Pretoria: Institute for Security Studies, ISS Monograph No. 20, 1998).

29. See, *inter alia*, M. Shaw, "Biting the bullet: negotiating democracy's defence" in *South African Review 7: The Small Miracle. South Africa's Negotiated Settlement*, ed. S. Friedman and D. Atkinson (Johannesburg: Ravan Press, 1994), 228-56.

30. The Military Research Group was coordinated by South African scholar Gavin Cawthra, based at the University of Witwatersrand in Johannesburg, and included other academics like Laurie Nathan, Jakkie Coq and Rocky Williams. It disbanded by about mid 1995. Author's interview with Gavin Cawthra, 22 June 2004.

31. See United Nations Development Program, "Human Development Report, 1994," http://hdr.undp.org/reprts/global/1994/en, (25 March 2005).

32. See, *inter alia*, Jakkie Cilliers and Hussein Solomon, Editors, *People, Poverty and Peace: Human Security in Southern Africa* (Pretoria: Institute for Security Studies, ISS Monograph Number 4, 1996); and H. Solomon and M. van Aardt, Editors, *Caring Security in Africa* (Pretoria: Institute for Security Studies, ISS Monograph Number 20, 1998).

33. Greg Mills, "The South African National Defence Force: Between Downsizing and New Capabilities." *Naval War College Review* LII, no. 1 (Winter 1999): 79-80. The terminology here may be confusing to American readers. In the British and South African model, the "Secretary" in a cabinet ministry is not the head of the ministry, but the senior permanent civil servant. This position is designed to promote non-partisan functioning and continuity in the ministry as the political appointees in the most senior positions come and go.

34. Ibid., 79. Still, Heitman adds a cautionary note. The new arrangement preserved a considerable role in defense policymaking for the (uniformed) Chief of the South African National Defence Force, somewhat compromising the intended strength of civil control. Helmoed-Römer Heitman, personal communication with the author, 20 April 2005.

35. Ibid., 80-2. Author's interview with Len Le Roux, Director of the Defence Programme, Institute for Security Studies, Pretoria, 10 June 2005. See also http://www.info.gov.za/aboutgovt/parliament/committees.htm, 12 August 2005; http://www.pmg.org, (12 August 2005).

36. See, *inter alia*, Constitution of the Republic of South Africa, 1996 (Act No. 108 of 1996), particularly Chapter 11, paragraph 200: "The primary object of the defence force is to defend and protect the Republic, its territorial integrity and its people." http://www.info.gov.za/domuments/acts/index.htm#1996, (25 March 2005). Cawthra interview, 22 June 2004. See also Rocky Williams, "Defence in a democracy: the South African Defence Review and the redefinition of the parameters of the national defence debate," in *Ourselves to Know*, ed. R. Williams, G. Cawthra and D. Abrahams (Pretoria: Institute for Security Studies, 2003), 212. Williams reflects some of the diversity of the debate by taking issue with the "primacy" of the (military) prime function in his article.

37. Ibid.

38. Author's interviews with defense analyst Len Le Roux (a retired South African Air Force Major General) who served at senior level in the Defence Force headquarters during this period of transition and had regular contact with the actors he described), Pretoria, 11 March 2004; Gavin Cawthra, informal coordinator of the Military Research Group and now a scholar at the University of Witwatersrand, Johannesburg, 22 June

2004; and Helmoed-Römer Heitman, South African correspondent for *Jane's Defence Weekly*, 23 June 2004; information also derived from Heitman, personal communication with the author, 20 April 2005.

39. Ibid.

40. Batchelor and Willett attribute this at least partly to Armscor, which had remained within the Department of Defence and thus had privileged "insider" access where its personnel could lobby for a robust armaments production capability. Batchelor and Willett, *Disarmament and Defence*, 99, 155.

41. Kasrils was addressing a South African defense industry conference. Ronnie Kasrils, "The Future of South Africa's Defence Industry: the Government's Perspective, in *South Africa's Defence and Security in the 21st Century*, ed. W. Gutteridge (Brookfield, VT: Dartmouth Publishing Company, 1996), 117-26.

42. Heitman, personal communication, 20 April 2005. For detail, see Hilton Hamann, *Days of the Generals* (Cape Town: Zebra Press, 2001), 225-32.

43. For an interesting account of the role of business in the South African processes of political reform, see Michael Mann, "The Giant Stirs: South African Business in the Age of Reform," in *State, Resistance and Change in South Africa*, ed. P. Frankel, N. Pines, and M. Swilling (London: Croom Helm, 1988), 52-86.

44. Hamann, *Days of the Generals*, 209-23.

45. This entity ultimately settled on the title, Aerospace, Maritime and Defence (AMD).

46. Other industry figures that played important roles in the transition included the CEOs of Denel (Johan Alberts), Reunert (Boel Pretorious), Grintek (Sybrand Grobelaar) and ATE (Jean-Marc Pisano). Author's interviews with Reginald Ngwenya, Director of AMD, 21 June 2004, Alan Holloway, Vice Chairman of AMD and Managing Director of ANSYS, 21 June 2004 and Paul J. Gerber, Group Manager for Business Coordination and Counter-trade, Grintek Group Limited, 21 June 2004.

47. Interviews with various industry insiders, particularly Andre Nepgen, Director of CSIR/Defencetek, 20 June 2003.

48. Len Le Roux, "The South African National Defence Force and its involvement in the Defence Review process," in *Ourselves to Know*, ed. R. Williams, G. Cawthra and D. Abrahams (Pretoria: Institute for Security Studies, 2003), 155-6.

49. Based on the author's interviews with a wide range of South African scholars and former officials possessing first-hand experience in these events.

50. South Africans had widely differing expectations on almost any issue – including security – and to make matters worse, a relatively large number of government officials at all levels had little preparation for their new roles and responsibilities. But the new leaders of South Africa were determined to consult widely on major issues to avoid the abuses of the previous era. Cawthra, "South Africa," 142-3. See also, "Ready to Govern: ANC Policy Guideline for a Democratic South Africa," adopted at the National Conference, May 1992, http://www.anc.org.za/ancdocs/history/readyto.hmtl, (25 March 2005); Le Roux, "The South African National Defence Force," 154-63; and Mills, "The South African National Defence Force," 80-2.

51. Http://www.gov.za/documents, (25 March 2005).

52. Laurie Nathan, *The changing of the guard: armed forces and defence policy in a democratic South Africa* (Cape Town: HSRC Publishers, 1994).

53. See Le Roux, "The South African National Defence Force," 156-8.

54. Security was now an "all-encompassing condition in which *individual citizens*

live in freedom, peace and safety; participate fully in the process of governance; enjoy the protection of fundamental rights; have access to resources and the basic necessities of life; and inhabit an environment which is not detrimental to their well-being." [emphasis added] *White Paper on Defence*, 1996, 3 http://www.info.gov.za/documents/ white-papers/index.htm#1996, (25 March 2005).

55. The *White Paper* explicitly stated that the [armed forces] comprise "an important security instrument of last resort . . . [but not] the dominant security institution." Ibid., 3, 4.

56. Ibid., 4. Virtually the same ideas, in the same wording, had appeared in South Africa's Interim Constitution of 1993, and would be reflected in the final Constitution of 1996.

57. Ibid.

58. Ibid., Chapter 2.

59. Le Roux, "The South African National Defence Force," 159.

60. Ibid., 158-68.

61. *South African Defence Review*, 1998 (Pretoria: 1 Military Printing Regiment, 1998), vi.

62. Le Roux interview, 11 March 2004; Heitman interview, 23 June 2004.

63. Ibid; Le Roux, "The South African National Defence Force," 163-7. Thandi Modise and Zou Kota, both members of Parliament's Joint Standing Committee on Defence, were noteworthy contributors to these efforts. Information provided to the author in a private communication from Helmoed-Römer Heitman, May 2004.

64. *Defence Review*, 121.

65. Ibid., 129.

66. Ibid., 127.

67. Ibid., 133.

68. The counter trade system was implemented by the Cabinet in 1996. Helmoed-Römer Heitman, *Unpublished Manuscript* provided to the author in June 2003, 80.

69. Batchelor and Willet, *Disarmament and Defence*, 146-50.

70. Ibid.

71. According to the document itself, the execution and management of the project was overseen by a steering committee appointed by the NCACC. Represented on this committee were the Defence Secretariat, the Department of Trade and Industry, Armscor, Denel, the South African Aerospace, Maritime and Defence Industries Association (AMD), the Human Sciences Research Council (HSRC), the Council for Scientific and Industrial Research (CSIR), the University of Cape Town's Centre for Conflict Resolution and the University of Witwatersrand Graduate School of Public and Development Management. *White Paper on South African Defence Related Industries*, Introduction, Paragraph 11, http://www.info.gov.za/documents/whitepapers/index.htm#1999, (25 March 2005).

72. Cawthra, interview, 22 June 2004; Heitman interview, 23 June 2004. Batchelor, the paper's principal drafter was at the time an economist associated with the University of Witwatersrand. In 2005 he was a senior analyst employed by the United Nations.

73. *White Paper on South African Defence Related Industries*, Introduction, paragraph 5.

74. This it defines broadly as: "those clusters of organizations in the public and private sector, and commercial companies and business units of such organizations, which are directly or indirectly active in the provision of goods and services to security forces which are defined as armaments. This provision can include research, design,

development, production, assembly, test, evaluation, upgrading, procurement, export, import, maintenance, logistical support, human support or project management." Ibid. Paragraph 6. A more explicit description of these commercial entities is offered in Chapter 2, Paragraphs 39-44.

75. Ibid., Paragraph 10. This is stated in the context of South Africa's role as a responsible member of the international community.

76. Ibid., Introduction, Paragraph 8.

77. Ibid., Paragraph 9.

78. Ibid., paragraphs 9, 32-52 and 66. After recounting the history and surveying the status of South African "defence-related industries," the *White Paper* identified the special capabilities of these industries in the mid 1990s. It specified areas in which South Africa's industries exhibit a "core competency," or put another way, areas in which the South Africans saw themselves as having a special advantage. The areas are: electronics (further articulated as guidance and fire-control systems and avionics); weapons systems (including "weapons for aircraft, helicopters, ships, vehicles, artillery and infantry"); and communications (expanded to include secure communications, electronic warfare, radar and information technology). The paper identified six "strategically essential defence technologies and capabilities" that South Africa would seek to establish or maintain for its own security needs, whether or not they had further commercial viability. These technologies would maintain "independence in key strategic areas or certain niche areas [and reduce] supply vulnerabilities." The essential technologies are: logistic support, repair and maintenance of equipment and systems; systems integration (defined as "the ability to design, specify, procure, assemble and test the sub-systems, their interfaces and the resultant major system"); command, control and communications systems; sensors, signal processing and data processing; combat systems software and support (emphasizing cyber applications); and simulation systems and war gaming.

79. Ibid., paragraphs 53-61.

80. Ibid., paragraph 62.

81. Ibid., paragraphs 63-5.

82. Cawthra interview, 22 June 2004.

83. Leon Englebrecht, "Developments in the South African Defence Industry: International Alliances," summary of NIC/IISS Conference, "Transformation in Global Defence Markets and Industries," 4-5 November 2000, http://sss.armsdeal-vpo.co.za/special_items/reading/developments_sa_defence.html, (23 April 2004).

84. Particularly in Chapters 3, 4 and 8.

85. Particularly in Chapters 4, 5, 6 and 7.

86. Particularly in Chapters 2 and 3.

87. Particularly Chapter 4.

88. *South African National Defence Force Military Strategy*, (Pretoria: 1 Military Training Regiment, November 2002).

89. Ibid., 4.

90. According to Heitman, in 2005 the South African National Defence Force contingent in Burundi equated to a "Battalion Group" (in this case a ground force battalion augmented with VIP protection and other elements). The South African deployment to the Democratic Republic of the Congo included a Battalion Group (reinforced battalion) along with separate Air Force and Medical Service assets. Heitman, personal communication, 20 April 2005. For a useful overview of these missions, see Vanessa Kent and

Mark Malan, *Decisions, Decisions: South Africa's foray into regional peace operations* (Pretoria: Institute for Security Studies, ISS Occasional Paper 72, April 2003). For an assessment of the South African role *vis a vis* the Democratic Republic of the Congo, see particularly Lt.Col. Naison Ngoma (Zambian Army, retired), *Hawks, Doves or Penguins? A critical review of the SADC military intervention in the DRC* (Pretoria: Institute for Security Studies, ISS Paper 88, April 2004).

91. *White Paper on Defence*, 19. *White Paper on South African Participation in International Peace Missions*, available from http:www.trainingforpeace.org/events/archive/whitepaper.pdf.

92. South African scholars like Rocky Williams and Peter Batchelor have called attention to this apparent contradiction. It also was publicly recognized in 2003 by the Commanding Officer of the South African Army who implicitly criticized the limitations of the *White Paper on Defence*. See Batchelor and Willett, *Disarmament and Defence*, 135-7; Rocky Williams, "Defence in a democracy," 205-23; and S. Nyanda, "The South African National Defence Force and Peace Missions in Africa," paper presented at the African Dialogue Series, University of Pretoria, 27 February 2003.

93. Heitman, personal communication with the author, 12 June 2004.

94. Heitman, personal communication, 20 April 2005; Wyndham Hartley, "African Peace Burden Cannot be SA's Alone Warns Lekota," (Johannesburg) *Business Day*, 16 February 2005.

95. Heitman, *Unpublished Manuscript*, 62.

96. That committee includes the Ministers of Foreign Affairs, Defence, Intelligence, Safety and Security, Home Affairs and Correctional Services and Justice. The "Strategic Packages" announced in late 1998 are the responsibility of a Cabinet committee comprised of the Ministers of Defence, Finance, Trade and Industry and Public Enterprises. Heitman, *Unpublished Manuscript*, 11. Heitman observes that detailed consideration of defense policy devolves upon Ministerial sub-committees comprised of the cabinet ministers concerned with particular issues and chaired by the Minister of Defence.

97. South African Constitution of 1996, Article 228(3). See also *White Paper on Defence*, Chapter 3, paragraph 4; and Mills, "The South African National Defence Force," 80-2.

98. Heitman, *Unpublished Manuscript*, 12. See also *White Paper on Defence*, Chapter 3, paragraph 4; and http://www.info.gov.za/aboutgovt/parliament/committees.htm, (25 August 2005).

99. The composition of the Defense Secretariat emphasizes civilian supremacy in defense policymaking. In 2004 it was composed of five divisions: Policy and Planning, Finance, Acquisition and Procurement, Inspector General and Affirmative Action/Equal Opportunity. Two of these are particularly influential in acquisition: the Policy and Plans Division is responsible for "developing defence policy within the guidelines set out by the Cabinet, and then translating approved policy into a long-term departmental plan," while the acquisition and Procurement Division develops policy regarding the defense industry. Neither of these is particularly unique in a parliamentary democracy, but it is worth calling attention to another – the Affirmative Action/Equal Opportunity division – intended to redress some of the effects of discrimination in South Africa's recent past. As evident in the concerns of this division, government personnel appointments and contracts awarded to private sector enterprises are two of the many tools now employed by South Africa to empower its economically disadvantaged citizens. Heitman, *Unpublished Manuscript*, 11.

100. Armscor's *apartheid* era function was first comprehensively defined in 1968 legislation (Armaments Development and Production Act, 1968, Act No. 57 of 1968) and was adjusted by various other Acts and then by administrative fiat in the early 1990s. But it was only in 2004 that it received a wholly new legislative mandate (Armaments Corporation of South Africa, Limited Act, 2003, adopted in April 2004), repealing the foundational 1968 Act and all or part of eight others. Based on the new legislation, Armscor's primary role still is the supervision of defense acquisition. See also *Defense Review,* Chapter 13, Paragraph 31; *White Paper on South African Defence Related Industries,* Chapter 2, Paragraphs 31-8.

101. *White Paper on South African Defence Related Industries,* section 3. See also *Defence Review,* Chapter 13, Paragraph 36.

102. This is based on its mandated function in the *White Paper on South African Defence Related Industries,* Chapter 2, Paragraphs 32 and 36.

103. One South African informant characterized the overall arms industry and arms acquisition process as a "clock" in which all the cogs and gears were moving, new ones are added from time to time, but few seemed really connected to the others. Another informant contemptuously dismissed the large Armscor contingents that allegedly accompany official delegations abroad, enjoy many perks, and yet rarely find significant marketing opportunities. These perspectives were offered during the author's interviews with mid-level industry insiders in 2003 and 2004.

104. Heitman interview, 23 June 2004.

105. Http://www.amd.org.za, (12 September 2004); Ngwenya interview, 21 June 2004; Holloway interview, 21 June 2004; Gerber interview, 21 June 2004; Heitman interview, 23 June 2004.

106. This information is based on the comments of numerous mid-level South African arms industry insiders in conversations with the author in 2003 and 2004.

107. This new organization essentially performs the function of a Marketing Support Board proposed in Chapter 5 of the *White Paper on South African Defence-related Industries.* For additional information, see Botha, *South Africa's Defence Industry,* 4; Armscor *Annual Report,* 2003-2004, 43, http://www.armscor.co.za, (25 March 2005) and the Armscor promotional literature at http://www.sadid.co.za/SADID_7/edition7/armscor/armscor.html, (15 September 2004).

108. Information provided in personal communication by Paul J. Gerber, Group Manager for Business Coordination and Counter-trade, Grintek Group Limited, 22 April 2004.

109. The terminology here may be confusing to American readers. In the British and South African model, the "Secretary" in a cabinet ministry is not the head of the ministry, but the senior permanent civil servant. This position is designed to promote non-partisan functioning and continuity in the ministry as the political appointees in the most senior positions come and go.

110. The Board of Directors includes seven AMD nominees, along with senior representatives from the Departments of Foreign Affairs, Trade and Industry, Science and Technology and the South African Secret Service. The Executive Committee for day-today operations draws personnel from the same sources. Information provided by Paul J. Gerber, Group Manager for Business Coordination and Counter-trade, Grintek Group Limited, 22 April 2004.

111. In 2005, the most visible example of this issue was the suit brought by South African businessman Richard Young, proprietor of the small information technology firm

$C^2 I^2$ in Rondebosch. Young's suit alleged improprieties in the tender process by which Armscor awarded a significant defense contract to the French-owned multinational African Defence Systems (ADS), whose South African subsidiary is partly owned by the brother of former Department of Defence acquisition chief Shamir ("Chippy") Shaik. For additional detail, see Tim Cohen and Nicola Jenvey, "Court gives go-ahead for R150m arms claim," (Johannesburg) *Business Day*, 31 January 2005

112. Nic Dawes, "No plane, no gain?" *Mail and Guardian*, 9 December 2004; Alan Holloway, Vice Chairman of AMD and Managing Director, ANSYS, personal communication with the author, 4 April 2005.

113. Author's interviews with, *inter alia*, Ngwenya, Holloway and Gerber, 21 June 2004; Cawthra, 22 June 2004; van Rensburg, 23 June 2004 and Heitman, 23 June 2004.

114. Ibid.

CHAPTER 3

Defense "Packages," Counter-trade "Offsets" and Foreign Connections

A New Context

Between 1960 and 1985, an increasingly sophisticated and prolific arms industry took shape in South Africa, a country whose ruling white minority was motivated by a profound sense of insecurity and a perceived need for ever greater military capability. The growth of the industry was stimulated in part by UN sanctions on arms sales to South Africa. The advent of the UN sanctions, first in 1963 and then in 1977, progressively limited the country's access to the international arms market, while conflict in Angola and unrest at home in the mid 1970s reaffirmed the worst fears of its leaders. White South Africans increasingly saw themselves alone in a desperate struggle against an international communist menace, lavishly supplied from Soviet military arsenals and bent on seizing the country by subversion and naked force. These fears motivated them to seek self-sufficiency in armament production, and while they never fully achieved it, they came remarkably close for a developing nation.

South Africa's leaders also had concluded by the late 1970s that they could reap significant economic benefit through exports of their increasingly sophisticated, combat-proven materiel. In the 1980s, their weaponry began to appear in international arms expositions, and despite a 1984 UN Security Council Resolution against purchases of such materiel, the South Africans achieved significant export success, particularly in sales of artillery and artillery ammunition. By the late 1980s arms exports had become an important feature of the economy. Still, despite the very real economic benefits, neither the exports nor the local consumption of arms could compensate for the structural anomalies in the national economy produced by the massive investment in the arms industry. In the immediate aftermath of majority rule, maintaining that industry made much less sense economically or strategically.

Yet, as noted in the previous chapter, the South Africans did not discard their armaments production capacities along with the other baggage of the *ancien régime*. Instead, they engaged in a remarkable consultative process in the 1990s, in which they elected to adapt their arms industry to the strategic and economic priorities of the new South Africa. A warrant for the evolving relationships between government, industry, society and external partners was established by several key official documents, particularly the *Defence Review* (1998) and the *White Paper on the South African Defence Related Industries* (1999). But the more interesting aspects of the story are in the specific measures the South Africans undertook to implement their new thinking. This chapter examines three general areas in which the South Africans sought to promote both their security and their national economic well-being: the use of countertrade "offsets," the purchases of foreign weapons containing South African components, and the foreign partnerships that grew primarily as a result of the first two programs.

The South African "Industrial Participation Programs"

Like a majority of the world's nations, South Africa engages in countertrade relations with its international partners. In other words, the country requires reciprocal and compensatory agreements from external sellers that enable it to reduce the net cost of its foreign purchases. Since the mid 1990s, South Africa has required foreign purveyors of armament to enter into "offset" agreements – a particular form of countertrade.[1] These agreements are "commercial transactions in which the buyer demands, as a condition of sale, that the seller offer compensation through a variety of nonmonetary means."[2] The basic idea is that the seller "sweetens the deal" for the buyer by agreeing to investments or transactions in which the buyer (e.g. South Africa) reaps negotiated collateral economic benefits from the foreign trade partner.

Offset agreements generally can be categorized into one of two major varieties, direct and indirect. "Direct offsets" take advantage of the opportunities offered by the major purchase itself, and might for example allow the buyer to produce subassemblies or components under license. "Indirect offsets" involve the exchange of goods or services between the two parties that are not directly related to the original transaction. In contemporary practice, large-scale purchases by national governments typically feature a complex package that includes both varieties.[3] This also has been true of South Africa's arms purchases.

Offset arrangements have appeared on the international scene in an array of different forms including joint production by seller and purchaser, licensed production in the purchaser's country, exchange of agricultural or mineral commodities, or even involvement of the seller in the economy of the purchasing country by investment in local industries and development of enterprises such as tourism. These arrangements are an important feature of world trade. By the mid 1980s, almost ninety of the world's national governments – an overwhelming majority – required some form of countertrade in their business dealings.[4]

Weapons purchases typically have formed a substantial proportion of this world-wide commercial practice. Hammond's research suggested that by the mid 1980s, up to 20 percent of the world economy involved offsets derived from military counter-trade.[5] Conceivably, this percentage decreased in the wake of the Cold War as most of the competing parties no longer pursued arms races, but the precedent itself remains well established.

Recipient governments typically tout the prospective economic benefits of offset agreements, and such arrangements can provide substantial benefits to certain communities of constituents, but the overall benefits tend to be more political than economic. Economists debate the pros and cons and the objective economic value of offsets, and have been reluctant to endorse them wholeheartedly.[6]

When the South African government was considering an offset regime in the mid 1990s, it sought the counsel of international experts. The response apparently was not very encouraging – few other countries seem to have found such programs an unalloyed blessing. However, South Africa ultimately implemented its own – calling them the "industrial participation" programs – and the country has been shrewd in their use. Arguably, South Africa is one country for which offsets have proven generally beneficial.[7] South Africa actually has two separate offset protocols: the National Industrial Participation Programme (NIPP) managed by the Department of Trade and Industry and the Defence Industrial Participation Programme (DIPP) managed by Armscor on behalf of the Department of Defence.[8] The South African Cabinet implemented the first of these programs in late 1996.[9]

The NIPP requires a foreign supplier for any single South African government purchase exceeding US$10 million to invest at least 30 percent of the contract value in South Africa's economy. Arrangements for the investment must be finalized no later than seven years after completion of the basic contract. (This obligation rises to 50 percent in the case of defense contracts.) South Africa requires that NIPP investment be placed in enterprises other than the defense related industries. The intention behind the NIPP is that government purchases force external investment in South African industries, create linkages and technology transfers with external industries, minimize the outflow of currency reserves, create jobs (particularly among historically disadvantaged communities) and provide opportunities for export of "value added South African goods and services."[10] An Industrial Participation Secretariat within the Department of Trade and Industry, directed by a senior civil servant, emphasizes the importance of the NIPP.[11]

South Africa applies a second offset protocol, the DIPP (over and above the NIPP), to foreign suppliers of *defense*-related materiel in any single contract exceeding US$2 million. Unlike the NIPP, this program specifically requires the foreign seller to invest in the arms industry itself. It is designed to promote industry jobs, capabilities and marketing opportunities. The DIPP has several tiers of obligation, ranging from 50 percent to 100 percent of contract value.[12] Management of the DIPP is a joint responsibility of the Department of Trade and Industry and the Department of Defence. Armscor has a DIPP Secretariat to manage the Defence responsibilities. The Industrial Participation secretariats of both the Department of Trade and Industry and Armscor conduct separate evaluations and negotiations to arrive at DIPP agreements with foreign suppliers of defense-related materiel.[13]

The importance of these counter-trade offsets to South African industry cannot be overemphasized. Bitzinger argues that the DIPP provides more opportunity for the defense industry than exports: "offsets from [the arms 'packages' announced in 1998] could equal nearly 15% of the South African arms industry's total annual revenues for the next several years."[14] And this does not even consider the impact of counter-trade in South African economic development outside the defense sector.

The "industrial participation" programs have been a lucrative source of foreign investment for the country, but they have not been trouble-free. They have been a key feature of the "Defense Packages" announced in 1998, and those arms acquisitions have stirred controversy in South Africa, principally over financial improprieties in the purchases. Some of this discontent also derived from allegations of unfair preferences in the offset arrangements. Domestic disputes over the allocation of benefits have not been the only problems with offsets. Denel's financial director complained in late 2004 that his corporation had derived little profit from offset-driven contracts supporting the Defence Packages. And by early 2005, two of the foreign contractors involved in those Packages had publicly defaulted on their offset commitments.[15]

Still, the South African "Defence Packages" offer a graphic illustration of the impact and success of the country's offset policies. By late 2003, the successful foreign bidders had obligated themselves though the industrial participation programs to invest over four times as much in South Africa as the value of the defense-related materiel they were selling, a truly astonishing commitment.[16] It is worth examining the "Packages" in a bit more detail, noting their overlap with the South African offset programs.

The Defence "Packages"

By 1999, the *White Paper on Defence*, *Defense Review* and *White Paper on the South African Defence-Related Industries* had provided substantial definition to a new South African military establishment – and the South African armaments industry. But the armed forces themselves were in difficult circumstances as these policy frameworks were being adopted.[17] South Africa's military found itself facing painful shortfalls in funds for training, maintenance and equipment replacement. Mills, for instance, notes that between 1990 and 1998, the defense budget had fallen from 4.5 percent to a mere 1.6 percent of gross domestic product, a reduction of almost US$3 billion.[18] According to Cilliers, the Air Force in late 1997 had been obliged to halt flight activity except in cases of emergency and the Army lacked sufficient resources to continue normal training.[19] These problems partially were caused – and certainly exacerbated – by the enormous costs of integrating and rationalizing the human resources of the new military, and demobilizing redundant personnel in the wake of majority rule.[20] Whether as a result of government budget cuts or increased human resource expenses, the South African military establishment found itself in increasingly severe straits as the decade progressed. Both its senior military officers and its civilian management were clamoring for additional resources and new materiel.

As early as 1994, South Africa's military leaders had developed and publicized listings of critical needs, and several foreign countries were lobbying the government for the right to provide new materiel, interested particularly in establishing sole-source "strategic alliances."[21] The South Africans seem to have been surprised by the intensity of the foreign interest, and finally reacted in September 1997 by issuing requests to eleven countries for technical information on military equipment.[22] (Conspicuous by its absence on the list was the United States.[23]) South Africa's Department of Defence, National Defence Force and Armscor subsequently evaluated the responses.

By early 1998, the Department of Defence had offered the Cabinet a short-list of prospective suppliers for the equipment it considered most sorely needed.[24] The government had made its final selections in time to announce them at a November 1998 Defence Exhibition in South Africa, when it published its intention to acquire five distinct "packages" of military materiel.[25] The contracts for the packages were signed a year later, in December 1999. With a value at the time of US$3.9 billion they comprised the largest ever single arms purchase by South Africa.[26] The new equipment included:

- Twenty-eight Saab JAS-39 *Gripen* multi-role fighter aircraft manufactured by a BAE Systems/Saab Consortium in Sweden. South African military planners originally envisioned a fleet of forty-eight; however the total subsequently was scaled back to thirty-eight and ultimately to twenty-eight, a minimum deemed necessary for a viable force.[27] Delivery of these aircraft was scheduled for 2007-2011.[28] Completion of the first aircraft (in Sweden) was scheduled for late 2005, with subsequent flight-testing in Sweden prior to delivery.[29]

- Twenty-four British Aerospace Industries (BAE) KM 120 *Hawk* lead-in fighter-trainer aircraft from the United Kingdom. Delivery was scheduled for 2005-2009.[30] The first *Hawk* arrived in November 2003, was initially displayed in February 2004 and subsequently commenced flight tests at the Air Force's Overberg Test Flight and Development

Centre. The first two operational aircraft were scheduled for delivery in August 2005, with two per month delivered thereafter.[31]

- Thirty A-109 Light Utility Helicopters from the Italian-based Agusta-Westland. South African planners originally contemplated a fleet of sixty, but subsequent reevaluation reduced this to forty and then to thirty with an option to purchase another ten.[32] Of these, seven were to be built at the Agusta plant in Italy, the remaining twenty-three in South Africa. By early 2004, the first four had been assembled in Italy and were undergoing qualification testing, while the South African parastatal Denel was gearing up to produce the remainder of the A-109 fleet.[33]

- Four Meko A200 patrol corvettes manufactured by the German Frigate Consortium in Germany. South Africa retained the option to purchase a fifth ship, and (as of early 2005) seemed likely to exercise it at some point in the future. Three of the four corvettes had arrived in South Africa by mid 2004, though none of the ships was scheduled to be fully operational until about 2007.[34]

- Three Class 209 1400 patrol submarines manufactured by the German Submarine Consortium/Ferrostaal in Germany. South Africa retained an option for a fourth ship, but (as this is written in 2005) appeared unlikely to exercise it. The first boat was scheduled for delivery to South Africa in July 2005.[35]

- Four *Super LYNX* 300 maritime patrol helicopters produced by Westland Helicopters, UK. These were intended for service with the new patrol corvettes.[36]

The strategic packages originally had included new main battle tanks to replace the aging *Olifants*, but this acquisition was deferred, as was the "package" for the four Westland *Super LYNX* shipboard helicopters.[37] (The *Super LYNX* deal subsequently was reinstated rather quietly in late 2003, but it was unclear in 2005 if South Africa would ever revive the "tank package."[38])

The arms purchases were controversial at the time of their announcement in late 1998, and became even more so over the next several years.[39] Predictably, many South Africans found their government's enthusiasm for large-scale purchases of arms puzzling, given the apparent lack of an external threat and the long list of unmet domestic needs.[40] Nor were the reservations limited to the civilian populace. Some South African military personnel had deep reservations about the selection process and the materiel itself – questioning, among other things, the suitability of the Swedish-designed JAS-39 *Gripen* aircraft.[41] The selection of the BAE *Hawk* lead-in fighter-trainer also provoked continuing criticism, with media allegations that a competing model offered by the Italian firm Aermacchi offered superior performance and better pricing.[42]

By far the most significant elements of the controversy were continuing allegations of financial impropriety and graft.[43] Almost from the outset, these swirled around key Defence and Party officials involved in the decisions.[44] As early as 2001, they resulted in the arrest of ANC parliamentary party whip Tony Yengeni on charges of corruption and fraud.[45] By 2003, allegations of corruption centered around South African Deputy President Jacob Zuma, resulting in the arrest of his financial advisor. Two years later, Zuma's advisor was convicted and sentenced to a long term in prison. The criminal trial disclosed the likelihood that Deputy President had accepted a substantial bribe for his patronage, and in June 2005 Zuma himself was sacked by South African President Thabo Mbeki, a move that generated substantial controversy within the country.[46]

Suspicions of graft over the arms deal probably are inevitable given the large sums of money and the intensity of foreign and domestic interest in lucrative contracts, but the suspicions seem also to have had some basis in fact.[47] Still, with the increasingly lonely exception of individuals like perennial anti-arms activist Terry Crawford-Browne, South Africans by 2005 generally had accepted the arms deals as a *fait accompli*.[48]

One feature of the arms packages that initially tended to be lost in the controversies was the extensive offset advantages South Africa reaped from its purchases, with benefits both to the defense industry and to other South African economic enterprises. Many of these came directly from offset agreements. The government required offsets at least equal in value to the purchase costs of the foreign materiel it was buying; and in fact, the competition for South Africa's business was so fierce that South Africa used the value of the offers as one key mechanism to discriminate among the bidders. The DIPP offsets alone ultimately amounted to about sixty percent of contract value of the foreign arms, and the NIPP offsets an astonishing three hundred and fifty percent. The total value of the offsets was some US$16.4 billion, vastly exceeding what South Africa had committed to pay for the weapons themselves.[49] The Saab/BAE contracts alone were characterized as providing "proportionately . . . the largest offset clause in defence industry history.[50] Botha puts these sums in perspective by noting that 1999 South African spending on recapitalization of its Defence Force was a mere US$225 million.[51]

The growing impact of the Defence Packages on the South African economy was clearly evident by mid 2003, when some forty-five NIPP projects already had gained South African government approval, carrying a prospective value of US$6 billion.[52] The offsets had begun to pay substantial dividends to the economy in general, and investments by the foreign arms producers may have been approaching the point where the economy's most lucrative near-term investment possibilities were saturated.[53] But it was the South African defense industry that was deriving the greatest long-term benefit from the policy. It seems likely the most valuable feature of the Packages has been the subsidization of the defense industry rather than any particular increase in near-term military capability or economic growth outside the defense sector.[54] The benefits can be categorized in several different ways.

In each of the "packages," the extensive use of South African subcontractors and South African-produced components by the foreign arms producers provides an immediate, tangible investment in South Africa's armaments industry. The subcontracts span a broad spectrum, including (among others) the following:[55]

- For the German-built corvettes
  - Assembly of the main diesel engines in South Africa;
  - Installation of the South African designed and built *Umkhonto* ("Spear") surface-to-air missile system;
  - Installation of South African-built or upgraded gun systems (76mm and 35mm);[56]
  - Installation of South African electronic warfare equipment;
  - Installation of a largely South African-designed and built combat management system, communications suite and switchboards; and
  - Installation of South African-designed and manufactured optronic and radar trackers.[57]

- For the German-built submarines
  - Installation of South African electrical systems, including electronic warfare equipment;
  - Installation of a South African designed and manufactured periscope; and
  - Installation of South African-produced electronic warfare equipment.[58]

- For the Italian-designed A-109 Light Utility Helicopters
  - Production of most of the helicopters in South Africa under license;
  - Local production of airframes for export;[59]
  - Installation of a South African developed internal diagnostic system; and
  - Repair and maintenance of the (Turbomeca Makila) aircraft engines in South Africa.

- For the British-designed KM 120 *Hawk* lead-in fighter-trainer aircraft
  - Final assembly and testing in South Africa;
  - Installation of a South African developed internal diagnostic system;[60]
  - Installation of various avionics and internal parts produced in South Africa;[61]
  - Local manufacture of tailplanes and other components for all future BAE *Hawks*;[62] and
  - Local manufacture of the flight simulator.[63]

- For the Swedish-built JAS-39 *Gripen* aircraft
  - Power supply, avionics, communications and electronic warfare systems developed and produced in South Africa;[64]
  - The helmet-sight system designed and produced in South Africa;[65]
  - NATO standard pylons produced in South Africa; and
  - The main undercarriage-bearing section of the fuselage along with the aft section of the fuselage all produced in South Africa.[66]

These subcontracts alone were worth more than US$200 million for South African producers in 2003 prices.[67] But the direct infusion of funding for the Defence Package weapons is overshadowed by a host of additional indirect benefits. These include a substantial DIPP investment by the foreign arms producers in South African defense industries. In some cases this has resulted in purchase of shares in South African defense companies. In other cases, it has featured the outright acquisition of South African enterprises by European-based defense conglomerates. The benefits extend to the transfer of personnel and expertise between South African and European firms, significantly facilitating two-way movements of defense technology. The advantages include access to otherwise inaccessible markets for South African armaments, and a visibility on the international defense scene that otherwise would not have been possible.

The combination of counter-trade offsets and Defense Packages could be taken as evidence of a deliberate strategy in which South African policymakers carefully considered a spectrum of national interests and, contrary to popular desires, deliberately selected and coordinated these programs to pursue interests they considered most pressing. Heitman, who cites his close contacts with senior policymakers (including former Minister of Defence Modise) supports this assessment. He argues that the Department of Defence (and the South African Cabinet) selected the specific end items in the "packages" from a carefully vetted Armscor shortlist, using consistent criteria. Those included the prospects for building long-term strategic alliances with South Africa's major trading partners, the opportunity to establish long-term industrial connections between South Africa's arms industry and successful foreign arms conglomerates, the amounts of NIPPS and DIPPS offers, and (finally) the actual cost of the materiel. Seen through the lens of these criteria, the South Africans obtained some excellent bargains in negotiating their defence packages, leading Heitman to argue that the acquisitions made both economic and military sense.[68]

Still, critics of the packages probably also should be given some credence. South African defense strategists may or may not have seen their strategic situation with appropriate clarity. Personal venality and organizational parochialisms undoubtedly infused some of the processes both in selecting the Defence Packages and in allocating their domestic benefits. And it is true enough that the foreign jet fighters, corvettes and submarines may have limited usefulness for the kinds of missions that will engage the Defence Force for the foreseeable future. But decision-making in liberal democracies generally is about balancing interests and accommodating constituents. It also is about providing insurance for a future that can never be fully anticipated. It rarely results in the most "efficient" or technically rational solutions even in the best of circumstances.

What is increasingly clear is that the potential long-term benefits of the "Packages" to South African industry are substantial. These benefits probably more than compensate in the long term for any compromise in military capability or suitability of the specific equipment to South Africa's current defense requirements. Even in conceding points to the critics, it is not hard to make a case for the military merits of the packages. Heitman probably deserves the last word, when he challenges critics to submit concrete proof that the equipment actually selected is not suited to the general needs of the South African National Defence Force, or that any of the competing alternative systems would have provided South Africa a demonstrably superior military capability.[69]

The Defence Packages highlight, as perhaps nothing else does, the current status of the South African armaments industries. These industries, typically in partnership with European producers, have surged onto the international scene with products marketed both to "high-end" and "low-end" consumers around the world. It is worth a brief look at these partnerships.

The Foreign Partnerships

South Africa's connections to external defense industries are not new. During the early years of the *apartheid* era, South Africa produced arms under license from industries in Belgium, France and Italy. And even after the "mandatory" UN embargo of 1977, South Africa continued a close cooperation with Israel on arms-related projects. In time, however, all of these relations were soured by the international opprobrium that attached to South Africa's domestic policies. At the end of the *apartheid* era, foreign partners were paying a high political price for real or suspected relations with South Africa. The advent of majority rule, accompanied by an end of UN sanctions, vastly increased the country's options and by the beginning of the 21st Century, South African enterprises had entered into arms industry linkages with other nations that would have been unthinkable a decade earlier.

Some of the new partnerships were a result of the Defence Packages and the related Industrial Participation Programs described earlier in the chapter. The South African arms purchases necessarily brought foreign firms and consortia (Saab, BAE, Agusta, Westland, Ferrostaal and others) into cooperative arrangements with South African defense contractors. Beyond the Package awardees themselves, the prospect of South African arms purchases had induced other foreign arms producers (like the French-based Thales) to examine the possibility of connections with South African firms.

Some partnerships seem to have coalesced independently of the Defence Packages. For instance, the Swedish electronic warfare firm Celsius, later absorbed by Saab, invested in the South African Grintek in the 1990s. Grintek, for its part, also developed a partnership with the German firm Ewation (a subsidiary of the European aerospace giant EADS) largely because of opportunity and compatibility. Somewhat the same could be said of the British military vehicle firms Vickers and Alvis, which had largely absorbed the South African military vehicle industry by the time the British-based BAE bought it all out in 2004. (As this is written in early 2005, the South African military vehicle subsidiary is known as BAE Land Systems OMC.[70])

The advantages in South Africa's partnerships with European-based defense conglomerates have by no means all been one-sided. The European enterprises generally have better market access and more resources, but the South Africans have advantages of their own: in 2003 they dissolved a promising partnership between (the South African) Advanced Technologies and Engineering (ATE) and BAE, involving a 20 percent BAE ownership of the South African firm. The issue was a competition between the two firms over contracts to upgrade eastern European military aircraft. The South Africans believed (apparently with reason) they could offer better equipment, superior service and more competitive terms.[71]

South Africa's significant foreign arms partnerships have emphasized European-based industries for obvious historical reasons. In the early 21st Century, thousands of citizens of European countries maintained their residence in South Africa. And beyond the resident foreigners, the country itself had a significant community of European descent, some of which continued to value familial connections to ancestral communities in Europe. Many South Africans of all races have lived or studied in Europe. Beyond the ethnic and cultural connections, much of South Africa's military materiel is European designed, produced or licensed and thus woven into the fabric of the local military culture. Arguably, much contemporary European military technology is state-of-the-art and thus attractive on its own merits. But since the advent of majority rule, the European arms industries also have vigorously courted the South Africans, offering them technology, partnership opportunities and access to markets in the rest of the world. These features all contribute to a natural South African inclination to favor European defense industries.

While the South Africans value their relations with the Europeans, they have (since the advent of majority rule) also been ideologically inclined to seek out industrial and trade relationships with the developing world. They have explored the prospects for connections with Brazil and India. Since 2003, the South Africans have been engaged in a protracted series of discussions with those two countries about a wide range of potential commercial relationships, including coordination of defense industry production and marketing.[72] Each of the three countries has a significant arms industry and each has unique (and possibly complementary) niche capabilities and interests. The South African expertise in rotary wing aircraft production may complement Brazil's manufacture of fixed-wing aircraft. South Africa's sophisticated gun turrets conceivably could be compatible with India's expertise in tracked military vehicles.

While all sides of the India-Brazil-South Africa (IBSA) dialog seem to see potential in cooperative ventures, substantive agreements have so far proven elusive, hampered by a host of technical difficulties. Still, if the multilateral consultations ultimately bear fruit, the three countries may in the future offer jointly produced or complementary military systems that reflect the strengths of the different national industries. They may be able to take unique advantage of markets that are less accessible or attractive to the industries of the developed world.[73]

The US Connection

South African arms industry relations with the United States in the decade following majority rule were strangely ambivalent, mirroring in some degree the ups and downs of the bilateral political relationship. A handful of South African defense firms established some kind of partnership with American counterparts, and South Africa's arms producers hardly could afford to ignore the technology or the marketing opportunities offered by the world's largest economy. However, a wider US-South African cooperation on arms was impeded by a number of issues, reflecting the unfortunate baggage of the *apartheid* era and irritants of more recent origin.

In 2005 some senior members of both the South African government and industry still had profoundly negative views of the United States, based on lingering resentments of US foreign policy during the *apartheid* years. In an ironic twist, some of the older South African political actors had antipathies stemming from seemingly opposite early experiences. Many of the country's contemporary senior political leaders spent their formative years in the anti-*apartheid* struggle, and some had long-standing exposure to East Bloc political thinking that portrayed the United States as the chief enemy of "progressive" movements throughout the world, including those in southern Africa. On the other hand, many South Africans that served in the government, military or defense industry during the *apartheid* era believed strongly that their country had been locked in a desperate struggle against the Communist world during the Cold War, a bitter ideological conflict in which the United States failed as a stalwart champion. They saw the precipitous US abandonment of its African allies (including South Africa) in Angola in 1976 as a cynical superpower maneuver or an incomprehensible betrayal. The US withdrawal from the Angolan conflict in 1976 was almost immediately followed by advent of the Carter Administration (1977-1980), an American government more vociferously critical of South Africa than its predecessors, and the few remaining low-level bilateral military relationships were terminated. Added to this was the later egregious insult of the US Comprehensive Anti-Apartheid Act of 1986, which unambiguously condemned South Africa's political culture, undermined its economy and significantly encouraged its adversaries.[74] Some members of South Africa's current government and industry served either in the *apartheid* state or in the liberation groups fighting against it, and among these were individuals in 2005 that harbored continuing suspicions and bitterness towards the United States from their earlier experiences.

The resentments might largely have dissipated in the 1990s were it not for a series of issues that continued to trouble security-related relations between the two countries even after the South Africans had begun a concerted movement toward political reform in 1989. A major US preoccupation in the 1980s and 1990s (giving all the appearance of the zeal of the recently converted) was a seemingly obsessive concern to punish earlier violation of US exports laws. As early as 1980, the United States had prosecuted and briefly jailed Canadian-American arms designer Gerald Bull for assisting South Africa with the development of new artillery.[75] By 1989, the US Department of Justice had initiated a series of vigorous investigations into violations of arms sanctions. In the early 1990s, widespread rumors circulated in South Africa that US agents had kidnapped local industrialists and had transported them against their will to the United States for prosecution. Some South African industry leaders reportedly feared for their lives. While these concerns appear amusingly improbable in sober hindsight, they were taken very seriously in South Africa at the time and had an enduring influence on attitudes towards the United States.[76]

Despite South Africa's transition to majority rule in 1994, the government of the United States continued to pursue legal action against South African firms it accused of violating US law during the *apartheid* years. This resulted in the indictment in 1994 of Armscor (later including Denel), the Armscor/Denel subsidiary Kentron and the private sector enterprise Fuchs (a subsidiary of Reunert) for conspiring to evade US export control laws. The indictment alleged that the South Africans had from 1978 through 1989 established an extensive network of "front companies" to acquire US arms, munitions and technology without obtaining the necessary export permits. A civil suit brought by the US Department of State followed the criminal indictment. The two cases dragged on until South Africa and the United States reached a negotiated settlement to both cases in 1996. The indicted South African parties then signed a plea agreement (in 1997) that settled the criminal charges, but resulted in fines of $12.5 million and debarment of trade between the US and the firms. (The debarment was suspended in 1998, and fully rescinded in 2004.)[77]

The results of the legal action now seem remarkably mild in retrospect. No South African industrial leader was jailed or fired, and funds exacted by the fines were directed to the South African government to assist in creating mechanisms for oversight of arms exports. Still, many South African officials in business and government considered the US actions to be arbitrary and mean-spirited, particularly in light of the radical political changes taking place in South Africa after 1989.[78]

The prosecution of these cases poisoned the prospects for substantive arms industry relations between the two countries in the 1990s. Ironically, relations were at a particularly low ebb when South Africa was considering the military acquisitions later announced in the Defence Packages. By that time, European arms industries had developed substantial South African partnerships and were energetically soliciting South African business.

While officials in both the United States and South Africa were hopeful by 2005 that the resolution of the legal issues over export violations would facilitate normal trade relations, the arms industry connections still were significantly obstructed by US bureaucratic red tape left over from the earlier cases. The Department of State had fully rescinded the South African debarment, but the US Bureau of Alcohol, Tax and Firearms (BATF) had still not gotten around to amending its restrictions on imports of South African munitions.[79] In early 2005, this was impeding the import of items such as the South African artillery ammunition that the US Department of Defense was seeking to test.[80]

The prosecution of the arms industries in US courts was not the only issue that soured prospects for productive relations in the 1990s and beyond. South Africans also accused the United States of maliciously obstructing their opportunity to market military materiel to other customers abroad. Within the South African arms industry and military, a widely believed story (in 2003 and 2004) held that the Army of the United Kingdom was seriously contemplating the purchase of South Africa's *Rooivalk* combat helicopter in the late 1990s, when the United States refused to release a Raytheon-produced air-to-ground missile, whose presence in the helicopter was an essential condition of the sale. According to the story, this deliberately spiteful act prevented an otherwise enthusiastic United Kingdom from purchasing the South African helicopter. Like the fear of US kidnapping, this urban legend also had little basis in fact, but was no less influential in South Africa for its lack of empirical substance.[81]

Another issue that troubled security relations between the two countries (with some spin off for the arms industry) was the US donation in the late 1990s of older C-130 transport aircraft under the Excess Defense Articles (EDA) program.[82] The EDA program is intended to allow the United States to dispose of unneeded military materiel and at the same time assist foreign partners; however, the process can be problematic since a foreign recipient is obliged accept the equipment in whatever condition it happens to be when declared excess, and is fully responsible for its removal from the US-controlled location. These rules obliged the South African government to pay for the rehabilitation of the donated aircraft in order to make them airworthy, an expensive proposition in the case of large, complex machines like the C-130 – and even more so in the case of aircraft that are old, heavily used and drawn from long-term storage.[83] There was substantial grumbling in the South African military and defense industry about "substandard, surplus American military equipment" delivered at inflated prices, along with allegations that the C-130 "donation" was a scam at South Africa's expense whose beneficiaries were US businesses involved in aircraft renovation.

Regrettably, the C-130 issue was soon ensnarled in still another bilateral tiff. The United States had in 1994 initiated a security assistance program with South Africa under which it provided modest funding for military education and purchases of US military materiel. (In 1999, for instance, the US provided South Africa a total of some $1.3 million, just over $1 million for education and another $0.3 million for materiel.[84]) South Africa had begun to use its US security assistance grants to purchase spare parts for its US-made C-130 aircraft. And in fact, the US Congress allocated significant funds in fiscal years 2002 and 2003 to facilitate this usage – $6.7 million and $6.0 million respectively. However, in 2001, the US Congress also passed legislation prohibiting US security assistance to countries refusing to enter into so-called "Article 98" treaties with the United States (treaties that protected US military personnel from the threat of prosecution by the International Criminal Court).[85] South Africa found this requirement distasteful and resisted entering into such an agreement. As a result, US security assistance funding to South Africa had been suspended by 2004. This prevented the country from using funding already allocated to purchase spare parts, imposing a severe hardship on the economically strapped South African Air Force as it struggled to maintain its C-130 fleet.

Despite the various issues that have troubled bilateral relations, the United States and South Africa have continued to communicate on security issues and have gradually expanded the scope of mutual arms industry cooperation. A significant basis for this relationship was established soon after South Africa's transition to majority rule. In October 1994, US President Clinton and South African President Mandela agreed to create a Binational Commission (co-chaired by the vice presidents of each country) to identify prospects for collaboration. The first meeting of the Binational Commission was held in Washington in mid 1995. The Commission initially consisted of five committees, eventually growing to eight. One of the latter, established in July 1997, was the South Africa-US Defense Committee (DEFCOM).[86]

Within five years of its establishment, the Binational Commission as a whole was essentially defunct, but the DEFCOM continued to function and in 2005 seemed likely to endure into the foreseeable future. It has met annually, alternating between the US and South Africa. The DEFCOM's five subordinate bodies included an Acquisition and Technology Working Group (A&TWG) responsible for cooperation on issues of technology. This Working Group has facilitated a number of joint projects and is likely to oversee a modest but growing partnership in the future.[87]

In addition to the conferences of the DEFCOM and its working groups, since 1994 both the United States and South Africa had sponsored a variety of other low-level visits, exchanges and evaluations involving defense and arms industry officials. The US Army Program Manager for Close Combat Systems (PM CSS) had been particularly active, culminating in a significant 2005 purchase of South African countermine technology intended for situations like the insurgency in Iraq. Mid-level officers and technical experts in the US Air Force and the US Navy also have engaged in periodic discussions with their South African counterparts on technology of mutual interest.[88]

In 2005, the areas of actual US-South African cooperation on issues of military technology were concentrated primarily on ground force equipment. This included the testing (under US Army auspices) of advanced South African artillery ammunition and evaluation of new family of technologically sophisticated South African 105mm gun-howitzers.[89] It also featured the testing of South African protective systems for combat vehicles,[90] and South African humanitarian demining technology.[91] It included the examination of a new South African 20mm anti-materiel rifle.[92] By 2003, the DEFCOM's Acquisition and Technology Working Group had begun to explore the prospects for reciprocal access to ranges and test facilities in the US and South Africa,[93] and the Committee also had considered the possibility of employing the South African *Rooikat* combat reconnaissance vehicle in the highly realistic combat training environment of the US Army National Training Center at Fort Irwin, California.[94]

South Africa's mine-protected and environmentally protected vehicles have proven particularly attractive to US consumers. Land Systems OMC has been able to market its security vehicles (through General Dynamics Land Systems-Canada) to several law enforcement agencies in the United States and Canada and most recently to the US military. The US Special Operations Command also has tested this South African firm's light combat vehicles in Iraq and Afghanistan with promising results.[95] The vehicles provide a relatively high degree of protection from the improvised explosive devices widely used by Iraqi insurgents and have earned the respect of US personnel that have tested them.[96]

The most spectacular South African success in the countermine market was announced in February 2005, when the United States Army agreed to the US $38.2 million purchase of 148 South African mine-protected vehicles manufactured by Land Systems OMC (and marketed in North America by General Dynamics Land Systems-Canada). The South African firm expected to complete the entire order by the end of 2005 and was contracted to provide spare parts, technical documents, training and service back up.[97] The purchase itself was a result of growing awareness in the US military of its vulnerability to mines and improvised explosive devices. The conflict in Iraq spurred a significantly greater US military interest in countermine capability after 2003. By 2005 the US Army had developed plans for a significant expansion of its combat engineer resources to deal with mine threats, and these plans apparently will take extensive advantage of South African countermine technology.[98]

US-South African military cooperation in 2005 had not included the aerospace and electro-optical domains, where both countries maintain their most sophisticated technology. These also are areas of high security sensitivity by both countries, and a unilateral effort by either side to intensify cooperation probably runs the risk of ruffling political sensitivities. Still, in 2004 and 2005, South African industry insiders indicated to the author that their offers of South African military aerospace materiel for US evaluation had gone unheeded. Ironically, the US aviation industry (Boeing in particular) had not hesitated to invest in South African technology for commercial aviation components, and by 2005 was making substantial use of South African enterprises for civil aviation support services, appearing to be quite satisfied with the South African relationship.

One additional area of some US-South African cooperation has been the technology of low-intensity conflict, as seen in the US Army/CSIR demining projects, the US testing of Land Systems OMC light combat vehicles, and the US interest in the Denel (Mechem) 20mm anti-materiel rifle. The technology of low intensity conflict and complex humanitarian emergencies is probably an area where both sides could reap substantial benefits by intensified cooperation without much political downside. South Africa's military operations over the course of the latter half of the 20th Century and its substantial commitment to regional peace support operations in recent years (in Lesotho, Burundi and the Democratic Republic of the Congo) have highlighted the unique technological challenges at the low end of the conflict spectrum. It seems likely that the South Africans have the capability (and probably the motivation) to conceive, design and test a wide range of equipment, from uniforms and load-bearing individual soldier equipment specially adapted to the African environment to non-lethal weapons and tactical communications suited to the vast distances and dispersed locations typical of peacekeeping forces. These mutual interests could profitably be explored.

The foreign partnerships that have emerged since the advent of majority rule have provided South Africa's arms industry advantages and options that it simply did not have during the *apartheid* years. The book now turns to a somewhat closer look at the current nature and capabilities of that industry.

Notes

1. For a useful taxonomy and discussion of definitions, see Grant T. Hammond, *Countertrade, Offsets and Barter: An International Political Economy* (London: Pinter Publishers Ltd, 1990), 6-10. For the rich and growing literature on this topic, see "Offsets Information Website" at www.plymouth.ac.uk/politics.

2. Grant T. Hammond, "Offset, Arms and Innovation," *The Washington Quarterly* (Winter 1987): 175.

3. Ibid.

4. Hesham el-Abd and Michael Kenney O'Sullivan, "Encountering Countertrade," *Journal of Defense and Diplomacy* 2, no. 6 (1984): 23.

5. Hammond, *Countertrade, Offsets and Barter*, 11.

6. Peter Batchelor and Paul Dunne, "Industrial Participation, Investment and Growth: The Case of South Africa's Defence Related Industry," (Paper prepared for the Centre for Conflict Resolution, Cape Town and Middlesex University Business School, United Kingdom, February 2000), 4-6.

7. Ibid.

8. Details of the NIPP and DIPP protocols discussed here are taken from David Botha, *Offsetting the Costs of SA's Strategic Defence Packages* (Pretoria: Institute for Security Studies, ISS Occasional Paper 75, 2003); and Helmoed-Römer Heitman, *Unpublished Manuscript* provided to the author in June 2003, 80-3.

9. The National Industrial Participation Programme (NIPP) was the first of the two national programs. Botha notes that Armscor actually had required counter-trade offsets since 1989, though presumably as a more informal industry-to-industry arrangement than as a formal national program. Botha, *Offsetting the Costs*, 1.

10. Ibid.

11. Ibid., 2-3. An interdepartmental Industrial Participation Control Committee oversees the Secretariat and provides strategic guidance, develops policy and monitors compliance across the government.

12. Ibid., 3-5.

13. Ibid., 4.

14. Richard A. Bitzinger, *Towards a Brave New Arms Industry*? (New York: Oxford University Press, IISS Adelphi Paper 356, 2003), 46.

15. Denel Financial Director "Pottie" Potgeiter complained in November 2004 that Denel was "outmuscled in negotiation with European firms it had contracted to supply." Nic Dawes, "No plane, no gain? (Johannesburg) *Mail and Guardian*, 9 December 2004. The two foreign contractors that had publicly defaulted on their offset commitments were the German-based Ferrostaal and the French-owned Thales. Information provided by Alan Holloway, Vice Chairman of AMD and Managing Director of ANSYS, personal communication with the author, 4 April 2005.

16. Botha, *Offsetting the Costs*, 9.

17. Even as late as 2005, the South African military establishment still was struggling with resource issues stemming in large part from the personnel rationalization and affirmative action programs inherited from the agreements that mediated the transition to majority rule. See Wyndham Hartley, "Arms package faces critical risk as funds dry up," (Johannesburg) *Business Day*, 2 February 2005.

18. In 1997 US dollar values, this represented a reduction from $4.9 billion to $2.0 billion. Greg Mills, "The South African National Defence Force: Between Downsizing and New Capabilities," *Naval War College Review* LII, no.1 (Winter 1999): 87.

19. Jakkie Cilliers, *Defence Acquisitions-Unpacking the Package Deal* (Pretoria: Institute for Security Studies, ISS Occasional Paper 29, 1998).

20. The agreements that mediated the transition to majority rule in 1994 obliged the new South African government to integrate the 110,000 personnel in the armed forces of the old South African State (South African Defence Force) with 40,000 additional personnel. These came from two "liberation armies" (the ANC's *Mkhonto we Siwe* – MK, and the Pan African Congress' Azanian People's Liberation Army –APLA), the military establishments of four of the so-called South African "homelands" created by the *apartheid* state (Transkei, Bophuthatswana, Venda and Ciskei) and the Kwa Zulu Self Protection forces. The integration of the different military forces was to be followed by downsizing, rationalization and demobilization of excess personnel. This all proved to be an enormously slow, difficult and traumatic process of change, costly and politically fraught. For details, see Jakkie Cilliers, *The South African Army in Transition* (Pretoria: Institute for Security Studies, ISS Monograph Number 26, 1998). For discussion of the human costs of this process, see *inter alia*, Ian Liebenberg and Rialize Ferreira, "Loyal service and yet "demobbed" – Demobilization and the economic reintegration of South Africa's demobilized military personnel," *Journal of Asian and African Studies* 37, no. 3-5 (December 2002): 299-317; and Rialize Ferreira, "The integration of the armed forces into the South African National Defence Force: a sociological investigation," in *Armed Forces and the International Diversities*. Ed. Leena Parmar (Jaipur, India: Pointer Publi-

cations, 2001), 82-110.

21. Jakkie Cilliers, *Defence Acquisitions*, 3.

22. These were Brazil, Canada, the Czech Republic, Denmark, France, Germany, Italy, Russia, Spain, Sweden and the United Kingdom. All except Brazil and Denmark subsequently submitted bids. Mills, "The South African National Defence Force," 86.

23. South African defense analyst Len Le Roux (in an 11 March 2004 interview by the author) called attention to the poisoned relations on issues of arms trade between the United States and South Africa in the mid 1990s. This stemmed from the US indictment in the early 1990s of the South African firms Armscor (later including Denel) and Fuchs for conspiring to evade US export control laws in the 1970s and 1980s.

24. Cilliers, *Defence Acquisitions*.

25. Heitman, *Unpublished Manuscript*, 93.

26. David Botha, *Offsetting the Costs*, 1.

27. Information provided by Helmoed-Römer Heitman, personal communication with the author, 12 June 2004.

28. Ibid; Botha, *Offsetting the Costs*, 1. See also www.gripen.com. Flight trials for the first South African *Gripen* in Sweden were scheduled for 2005, with subsequent South African flight tests and equipment certification in 2006.

29. *ARMSCOR Annual Report, 2003-2004*, 17, http://www.armcor.co.za, (27 March 2005).

30. Botha, *Offsetting the Costs*, 5.

31. *ARMSCOR Annual Report, 2003-2004*, 17; Siseko Njobeni, "Air Force to Get First Two Hawk Fighter Trainers Next Month," (Johannesburg) *Business Day*, 6 July 2005.

32. Ibid; Heitman, personal communication, 12 June 2004.

33. *ARMSCOR Annual Report, 2003-2004*, 17.

34. Ibid., 18. The three corvettes delivered by early 2005 were the SAS *Amatola* (which arrived in Simons Town in November 2003), the SAS *Isandlwana* (which arrived in February 2004) and the SAS *Spioenkop* (which arrived in May 2004). Heitman, personal communication, 12 June 2004 and 20 April 2005; see also www.naval-technology.com/projects/meko and www.StrategyPage.com, (10 Nov 03). Missions for these ships include protection of marine resources, law enforcement, humanitarian operations in coastal areas, fire support for land forces, "military diplomacy," and peace operations.

35. Botha, *Offsetting the Costs*, 5; *ARMSCOR Annual Report, 2003-2004*, 18.

36. "South Africa Confirms *Super LYNX* 300 order," *Military Technology* 27, no. 10 (October 2003): 98; *ARMSCOR Annual Report, 2003-2004*, 17.

37. Heitman, *Unpublished Manuscript*, 93-4.

38. "SAFRICA: Defence Minister "Quietly" Signs Deal of Supply of Naval Helicopters," Global News Wire-Asia Africa Intelligence Wire, BBC Monitoring International Reports, 15 August 2003; "Lekota signs order for four naval helicopters, South African Press Association Global News Wire, 15 August 2003.

39. For a useful overview, see David Botha, "The Arms Deal Controversy," *African Security Review* 12, no. 3 (2003): 99-103.

40. "NUMSA Denies Arms Deal Payoff," *Cape Argus*, 1 December 1999; "$5.2bn military deal defended against backdrop of regional war," Financial Times Information, Southscan, 1 December 1998; see also http://www.ecaar.org/za/, (20 November 2004). Scholars like Gavin Cawthra implicitly criticized the purchases by arguing that expendi-

ture on defense – and particularly defense-related research and development – fails to provide the economic stimulus derived from investment in other sectors. Author's interview with Gavin Cawthra, Johannesburg, 22 June 2003.

41. This is based on author's conversations with approximately thirty South African active duty and retired military officers (ranging in rank from captain to major general) between 1999 and 2004. Some South African Air Force officers questioned the suitability of the *Gripen* for the southern African operational environment and the kinds of military operations they envisioned over the next several decades, although others pointed out that the criticisms reflected "older thinking" of the need for long range strike capability, not in line with the existing *Defence Review* of 1998. Defense analyst Len Le Roux (a retired Air Force Major General) and military correspondent Helmoed-Römer Heitman argue that the *Gripen* is relatively well suited to the defensive role envisioned in the *Review*. Le Roux interview, 11 March 2004; Heitman interview, 23 June 2004.

42. See, *inter alia*, "Obituary: Richard Charter," *Sunday Times*, 1 February 2004 and "26 January 2004: Newspaper continues anti-Hawk campaign despite official displeasure," http://www.saairforce.co.za/news.htm, (15 January 2005).

43. Botha, "The arms deal controversy," 102-3; Tim Cohen, "Arms-deal Report Cover-up," (Johannesburg) *Business Day*, 7 January 2005.

44. These have included, among others, the former Minister of Defence, Joe Modise, former Deputy Ministers Ronnie Kasrils and Max Sisulu, former Department of Defence Arms Procurement Chief Shamin Shaik (whose brother directed one of the beneficiary companies in South Africa) and even the governing ANC parliamentary party whip Tony Yengeni, who subsequently was successfully prosecuted for fraudulent dealings. "SADC Report Modiso's Dynamic Drawn into SA Arms Scandal" (Cape Town) *Financial Times*, 6 April 2001; "SA politics: Ambassadors quizzed over arms procurement kick-back claims," (Cape Town) *Financial Times*, 15 November 1999.

45. "National Report: Top ANC Official Arrested," *The Financial Gazette*, 4 October 2001.

46. Zuma is alleged to have solicited a large bribe from a defense contractor, and the evidence was sufficient to cost him his position as Deputy President. His adviser, Schabir Shaik (a brother of the former Chief of Acquisition in the Department of Defence, Shamin "Chippy" Shaik) was indicted in August 2003 on charges of corruption, fraud, theft, tax evasion, "reckless trading," and money laundering and convicted in June 2005. "Shaik in Court, Released on R1,000 Bail," SAPA, 25 August 2003; Robyn Chalmers and Hopewell Radebe, "Arms deal saga is sending wrong messages to the world," *Business Day* (South Africa), 29 August 2003; Anthony Stoppard, "Politics-South Africa: ANC Faces Corruption in its Ranks," *Global Information Network*, 30 August 2003 http://proquest.umi.com, (7 May 2004); Paddy Harper, "Shaik Letters Drop Zuma's Name," (Johannesburg) *Sunday Times*, 7 February 2005; Brendan Boyle, Dominic Mahlangu, S'thembiso Msomi and Sabelo Ndlangisa, "Quit or be fired" (Johannesburg) *Sunday Times*, 5 June 2005, 1; Paddy Harper, "Fall knocks the swagger out of Shaik," (Johannesburg) *Sunday Times*, 5 June 2005, 17.

47. Helmoed-Römer Heitman considers that much of the controversy was provoked and sustained by unsuccessful foreign bidders (and their South African supporters). He questions the motivation of the accusers and is skeptical of the merits of the accusations. Heitman, personal communication, 12 June 2004.

48. See, for instance, "Arms deal challenged in court," Sapa, 20 March 2003, available at http://iafrica.com/news/sa/220753.htm, (22 September 2004).

49. Botha, *Offsetting the Costs*, 5.

50. BAE promotional literature quoted by Botha, Ibid., 5.

51. Ibid.

52. Ibid. Botha points out that this sum represents fully forty percent of the US$14 billion NIPP commitments, a remarkable accomplishment achieved only three years after the contracts were signed. By late 2003, the NIPP projects undertaken by the foreign arms suppliers included:

- A joint venture for solar panels;
- Upgrading of ore smelters;
- Manufacture of medical waste containers;
- Manufacture of aluminum tubes for auto radiators;
- Manufacture of stamped metal panels for low-income housing;
- A joint venture for manufacture of gold chains;
- Textile manufacture;
- Manufacture and export of automobile leather seat covers;
- Manufacture of stainless steel;
- Floriculture export;
- Manufacture of high-quality condoms;
- Tourism marketing to European consumers;
- Biotechnology manufacture; and
- Forest management and product exploitation.

53. Ibid., 5 6.

54. In the author's interviews with a sample of about twenty mid-level defense industry insiders in mid 2003, none considered the Defense Packages to be a "bad deal" for South Africa. Their responses ranged from cautious optimism about future prospects to great enthusiasm – with more responses in the latter category than the former.

55. Botha, *Offsetting the Costs*, 6-7.

56. Heitman notes that the 76mm gun is the Italian-made Oto Melara *Compatto* removed from older South African Navy strike craft and refitted with a new South African electronic gun drive designed and built by Reutech. Heitman, personal communications, 12 June 2004 and 20 April 2005.

57. Ibid. Heitman calls attention to the fact that the stabilized platforms for the tracking equipment (built by Reutech) have been exported to Germany for installation on the new K 130 class corvettes, giving this South African technology a potentially global distribution.

58. Ibid. Heitman calls attention to the fact that by mid 2004, the South African EW systems had so impressed the German partners (and other clients) that they also been selected for installation on the German-made submarines being produced for Greece and South Korea. It appeared that the South Africa systems would be a popular feature in future sales by the German-based firm.

59. Ibid. According to Heitman, by mid 2004 South Africa already had a contract to build twenty A-109 air frames for the Swedish armed forces.

60. Ibid. Heitman notes that the South African-designed and produced HUMS

equipment now is standard on all *Hawks*, including those in Canadian and Australian service.

61. Ibid. According to Heitman, this includes the entire avionics suite, which was developed in South Africa.

62. Ibid. According to Heitman, this equipment would be installed on the *Hawks* purchased by the Royal Air Force in the future and those previously purchased that subsequently were programmed to undergo refurbishment.

63. Ibid. Heitman notes that the South African-designed and produced simulator is being produced for the Royal Australian Air Force.

64. Ibid. According to Heitman, this includes the microwave components of the radar.

65. Ibid. According to Heitman, the publicity for South African equipment gained from the *Gripen* contract contributed to the selection of the South African head position (helmet) sensor system for use in the Eurofighter *Typhoon* and Dassault *Rafale*.

66. Ibid.

67. Botha, *Offsetting the Costs*, 6-7.

68. Heitman, personal communication, 12 June 2004; interview, 23 June 2004.

69. Ibid.

70. The designation "OMC" recalls the *Olifant* Manufacturing Company, established in the 1970s to upgrade South Africa's aging British-made *Centurion* tanks into the South African *Olifant* ("elephant") version. OMC subsequently was acquired, sold and reacquired in series of divestments and consolidations. For additional detail, see Appendix B.

71. Author's interviews with Reginald Ngwenya, Director of AMD, 21 June 2004, Alan Holloway, Vice Chairman of AMD and Managing Director of ANSYS, 21 June 2004 and Paul J. Gerber, Group Manager for Business Coordination and Counter-trade, Grintek Group Limited, 21 June 2004.

72. The initiative first appeared with a meeting of foreign ministers in Brasilia in June 2003 and the respective heads of state in New York in September 2003. It took on a more formal aspect as the Trilateral Commission of the Ibsa (India-Brazil-South Africa) Dialogue Forum, with the first official meeting in New Delhi in March 2004. That meeting actually had been preceded in February 2004 by a meeting of the three Ministers of Defence in Pretoria. See http://www.dfa.gov.za/docs/2004/ibsa0305.htm, (26 March 2005). For related activity, see also Jonathan Katzenellenbogen and Wyndham Hartley, "SA, India and Brazil Enlist Private Sector Trade Plan," (Johannesburg) *Business Day*, 10 March 2005.

73. Ibid.

74. US President Reagan vetoed this legislation, reflecting his Administration's more conservative inclinations, but his veto was overturned by the US Congress.

75. The irony here is that while the South Africans financed Bull's development of the new artillery designs in 1975 and 1976, many South Africans believed – possibly with good reason – that the US Central Intelligence Agency also was supporting the development. For details of the life and work of Gerald Bull, see William Lowther, *Arms and the Man: Dr. Gerald Bull, Iraq and the Supergun* (Novato, CA: Presidio Press, 1991).

76. Author's interview of Johan van Rensburg, Commercial Specialist with the US Commercial Service, Johannesburg, 23 June 2004. The author has uncovered no evidence that US agents actually kidnapped any South African citizens for prosecution, or that US

authorities ever had seriously contemplated such action. However, given the strength of anti-*apartheid* resentments in South Africa in the mid 1990s, and uncertainties in the ongoing political transition in the country at the time, the South African industrialists probably could not be certain that their government would oppose the kidnapping – or that it even would object in any substantive way to such violations of South African sovereignty.

77. *Federal Register*, 63, no. 42 (4 March 1998), Notices, Public Notice 2752, Page 10671; *Federal Register*, 69, no. 139 (21 July 2004), Notices, Public Notice 4768, Page 43656.

78. Details provided by Johan van Rensburg, Commercial Specialist with the US Commercial Service, Johannesburg, personal communication with the author, 4 June 2004. This information also is derived from the author's multiple interviews with Len Le Roux, Director of the Defence Programme, Institute for Security Studies, Pretoria, June 2004 and June 2005.

79. The relevant BATF regulations are found at BATF Title 27, part 447.21. http://www.atf.gov, (18 March 2005).

80. Author's interview with Duncan Lang, contractor to the Office of the US Secretary of Defense, Alexandria, VA, 15 March 2005.

81. Helmoed-Römer Heitman corrects this account with a very different version: he says that the South African parastatal Denel sent an unsolicited *Rooivalk* to the United Kingdom, where it was thoroughly tested. Heitman agrees that British military pilots were very impressed with the performance of the aircraft, but says that the UK Ministry of Defence never seriously contemplated its purchase. Heitman suggests that the false story of a US connection might have been fabricated within Denel to justify earlier unwarranted assumptions that the United Kingdom would purchase the machine. Heitman interview, 23 June 2004.

82. Under this program, the US provided five C-130 aircraft to South Africa, three C-130f models and two C-130b models. These were allocated in Fiscal Year 1996 and delivered in 1997. Author's interview with Lieutenant Colonel James Miner, US Defense Security Cooperation Agency, 16 March 2004. See also http://www.dsca.mil/programs/eda/results.asp, (25 March 2005).

83. South Africa is not the only African country in which C-130 aircraft donated under the EDA program has caused significant bilateral tensions. Ethiopia, too, received the excess airplanes in the 1990s and its officials allegedly were stunned to receive a bill of several millions of dollars for aircraft rehabilitation.

84. The education funding is called International Military Education and Training (IMET). Grants for materiel are provided under the Foreign Military Financing (FMF) program. The IMET program provides the funding to allow foreign military students to study at US military schools. FMF is designed to allow foreign countries to purchase US military equipment with the US-provided funds. For details on security assistance funds for specific countries, years and programs, see http://www.fas.org, (25 March 2005).

85. In 1998, representatives from 160 countries met in Rome under UN auspices and agreed by treaty (now known as the "Rome Statute") to create an International Criminal Court (ICC). A US representative signed the treaty in 2000, but President Clinton subsequently refused to forward the treaty to the US Senate for ratification. In May 2002, the US formally notified the UN that it would not become a party to the Rome Statute. The treaty itself came into force in July 2002. Among other things, it provided a legal rationale and mechanism for prosecution of war criminals by the International Criminal Court.

However, Article 98 of the treaty allowed bilateral agreements between nations whereby each of the two contracting parties would refrain from surrendering their citizens for ICC prosecution except by mutual consent. In 2001, the US Congress enacted the American Servicemembers' Protection Act of 2001 prohibiting US security assistance to countries that are party to the Rome Statute unless they had entered into Article 98 agreements with the United States. By late 2004, almost 100 countries had done so; although a number of others refused, among them, South Africa.

86. This was established by agreement between then US Secretary of Defense Cohen and then South African Minister of Defence Modise.

87. Information provided by Duncan Lang, contractor for the Office of the US Secretary of Defense, March 2005. The four other DEFCOM working groups include: Policy, Military Relations, Security Assistance and Environmental Security.

88. Lang interview, 15 March 2005.

89. The advanced artillery ammunition is a joint development project involving General Dynamics Ordnance and Tactical Systems (GD-OTS) and South Africa's Denel (Naschem subsidiary). This project is designing longer-ranged and more lethal 105mm and 155mm artillery ammunition for the US Army, with initial operational capability expected in 2006. For its part, the new gun-howitzer is a joint project of General Dynamics Land Systems (GD-LS) and South Africa's Denel (LIW subsidiary). The project may ultimately develop a family of light 105mm gun-howitzers for use by the US Army and US Marine Corps based on the impressive new South African LEO (Light Experimental Ordnance) 105mm artillery system. This system conceivably could appear in a variety of roles. Its weight and size make it much more compatible with US requirements for expeditionary deployment than currently deployed systems. Information provided by Duncan Lang, contractor to the Office of the US Secretary of Defense, 12 March 2005.

90. Ibid. In 2005 the US Department of Defense, US Army and US Marine Corps were evaluating two related systems, developed by the South African private sector firm Avitronics (a division of Grinton, Pty, Ltd), for protecting ground forces vehicles. The first was a laser warning system that alerts a vehicle crew to threats such as an enemy range finder or beam-riding missile and initiates an automatic response by weapons on the vehicle. The second system was an active protection system that senses the threat, evaluates it, and automatically initiates countermeasures, including the launch of a variety of counter munitions.

91. Ibid. In the early years of the 21st Century, the US Army was managing four separate projects related to demining in collaboration with the South African Council for Scientific and Industrial Research (CSIR). These included development of protective mine boots (based on control of shock wave transfer), minefield marking, a VAMIDS Array Mount modification and a test project (conducted in Croatia) to compare the mine-detecting qualities of specially trained dogs (a specialty of the South African firm, Mechem) to current mechanical/chemical mine detection technology.

92. Ibid. Although the United States had not identified a specific requirement for such a weapon as of early 2005, the NTW-20 anti-materiel rifle produced by the South African firm Mechem (along with ammunition produced by Pretoria Metal Pressings – PMP) had so impressed officials of the US Army, US Marine Corps and US Special Operations Command that they were considering a formal evaluation of the weapon. With an effective range of 2,300 meters, versatile ammunition payload, ease of maintenance, and light weight (particularly when disassembled into two man-portable 33 kilogram backpacks) the weapon has obvious potential uses for the kinds of conflicts encountered

by the US military in Afghanistan and Iraq.

93. Executive Summary, Republic of South Africa-United States of America: 7th Annual Defense Committee Meeting: 4 to 6 June 2003, Annapolis Maryland, http://www.defense link.mil/policy/isa/africa/DEFCOM03minutes.htm, (21 March 2005).

94. Ibid.

95. Author's interview with Brigadier Tony Savides (South African Army, retired), Director of International Marketing, Alvis-OMC (now BAE Land Systems OMC), 18 June 2003.

96. Author's interview with Colonel Rex Forney, US Army, 18 April 2005. Forney is a Military Police officer who served in a senior leadership position in the US military in Iraq in 2004 and 2005 and supervised US forces employing the South African-made equipment. See also Sashni Pather, "How SA-made vehicle saved US troops," (Johannesburg) *Sunday Times*, 24 October 2004 http://www.sundaytimes.co.za, (28 October 2004); Michael Evans, "Deadly Rockets Blast Way Through," *London Times*, 10 November 2004; Cpl. Joe Niesen, "New vehicles helping against IEDs," ARNEWS, 11 February 2004 http://www4.army.mil, (28 October 2004).

97. The actual vehicle is the RG-31 medium mine-protected vehicle. At the time of the announcement, the purchase was described as "the largest single new vehicle export, by volume, ever placed with a South African defence manufacturer." See "SA firm in R468m deal to supply vehicles to US army," Creamer Media, 2005/02/25 http://www.mining weekly.co.za, (15 March 2005).

98. This expansion includes activation of seventeen new US Army Engineer Clearance Companies. As conceptualized in 2005, each of these new companies would field twenty-two South African RG-31 vehicles made by BAE Land Systems OMC and twelve IMMMD "Chubby" mine detection vehicles built by the South African firm RSD Dorbyl (along with other US-built equipment with South African technological antecedents). Lang interview, 15 March 2005.

CHAPTER 4

The South African Armaments Industry

General

South Africa's defense related industries now form a community that is much smaller and more externally oriented than it was at the end of the *apartheid* era. It also is a community more sophisticated and competitive, and more secure, having largely defined its role in the new South Africa. As has always been the case, much of the industry is concentrated in Gauteng Province, in the region around the two major population centers of Johannesburg and Tshwane (Pretoria). However, a surprising amount is found elsewhere, including particularly the Western Cape region. Employment and other direct and indirect economic benefits extend to a considerable proportion of the country.

Many of the old structures and organizational culture still are evident, but both have evolved in significant and novel ways since 1989. The government still coordinates their efforts to serve national priorities, a legacy of the years before majority rule, when the National Party government created an arms industry and then carefully controlled its activity to meet the security needs of the state. However, the motivation of the current African National Congress government is quite different. It admits to facing no immediate external military threat, yet seeks to safeguard defense industries both to preserve "strategic capabilities" and at the same time provide jobs for citizens and economic benefit to the state through arms sales and offset trade agreements. There is at least one other substantial benefit. As Heitman is quick to note, the industry is a good "incubator" for technology and provides an excellent environment in which to "grow" new generations of engineers, scientists and technicians.[1]

South Africa's arms producers still are an interesting combination of intertwined public and private sector entities. Those in the public sector now consist of three major actors: Armscor (still a component of the Department of Defense), Denel (a parastatal entity subordinate to the Department of Public Enterprises) and Defencetek, a quasi-public sector subsidiary of the Council for Scientific and Industrial Research (CSIR). For their part, the private sector industries are quite diverse with some of the arms producers imbedded in larger industrial groupings while others are autonomous enterprises. To understand South Africa's current arms production capability, it is useful to have a grasp of the diversity of the country's defense industry, starting with the public sector entities.

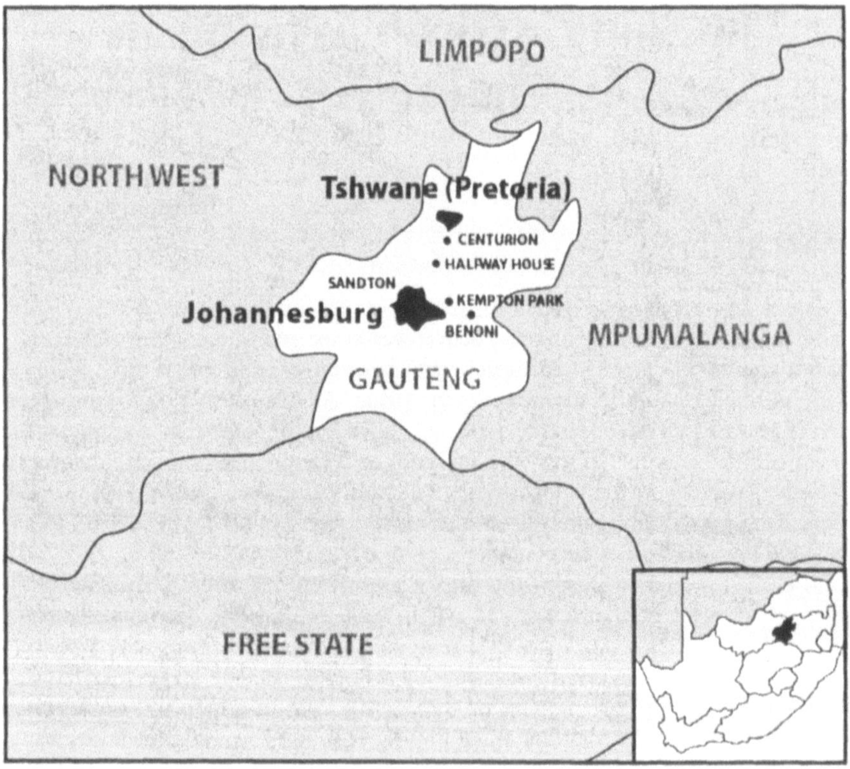

Figure 4.1. Key Industrial Towns in South Africa's Gauteng Province

Defense-related Industries in the Public Sector

The presence of Armscor and CSIR in South Africa's contemporary public sector is an obvious continuation of policies established early in the *apartheid* years, but today's Armscor is quite different in structure than it was in 1989. One of the key features in the public sector arms industry "devolution" of 1992 was the creation of Denel – the enterprise that assumed most of the arms production role formerly played by Armscor. And while Denel still was a public sector entity, it was transferred out of the Department of Defence, leaving only one of the three parastatal arms industry actors (Armscor) as a direct Defence subordinate. This sent an important signal about the importance of civilian control of the military in the new South Africa, and at the same time reflected some South African government ambivalence about the role of the country's arms production. Despite continuing assertions in official documents that the arms industry is key to preserving South Africa's strategic autonomy, the industry's role as a national money-maker and employer has become at least as important as any military capability it might bestow on the nation.

Armscor

While it has changed significantly in structure since 1992, Armscor remains an important arms industry actor in its own right. During the *apartheid* era, it became the massive, public-sector entity responsible for producing or acquiring all of the military materiel needed by the country's security establishment. It did not manufacture all of the country's arms, but made most of them and oversaw all acquisition and arms export processes. In these roles, it supervised its own subordinate enterprises for researching, testing and producing military materiel. It also served as the major contractor for armament purchased from the private sector or procured outside the country, a role that was clearly defined by legislation in 1968.[2] However, the separation of Denel from Armscor in 1992 significantly altered Armscor's role. Armscor retained the responsibility for procuring and testing the country's military materiel but gave up an explicit production function.

As has been the case since its founding in the 1960s, Armscor in 2005 remained a corporation wholly owned by the government of South Africa. By law, the Minister of Defence "exercises ownership . . . on behalf of the state."[3] An eleven-person board of directors chaired by an investment banker oversaw it.[4] The organization was headquartered in the large Armscor building complex in the Tshwane area, sharing these facilities with the headquarters of the South African National Defence Force. With its various subsidiaries, Armscor employed a total of about five thousand people. Its subordinate entities were located across South Africa (see Appendix A for details). It retained strong, continuing linkages to other defense industry actors in a variety of roles.

In 2005 Armscor's "core function" still was defense acquisition, a role it continued to play for the South African military establishment and other public sector security agencies.[5] Its promotional literature characterized its basic mission as "meet[ing] the [materiel] acquisition, maintenance and disposal needs of the South African Department of Defence and other clients."[6] Mandated functions included management and execution of Department of Defence research and development, control of the tender process for defense acquisition, execution of acquisition programs, disposition of defense materiel, management of strategic facilities, co-supervision of the South African defense industry as a whole (with the Department of Trade and Industry), marketing for the defense industry and management of the Defence Industrial Participation Program (DIPP) on behalf of the Department of Defence.[7] Not surprisingly, overseeing the "Strategic Packages" dominated Armcor's operations from 1999 through 2005.[8]

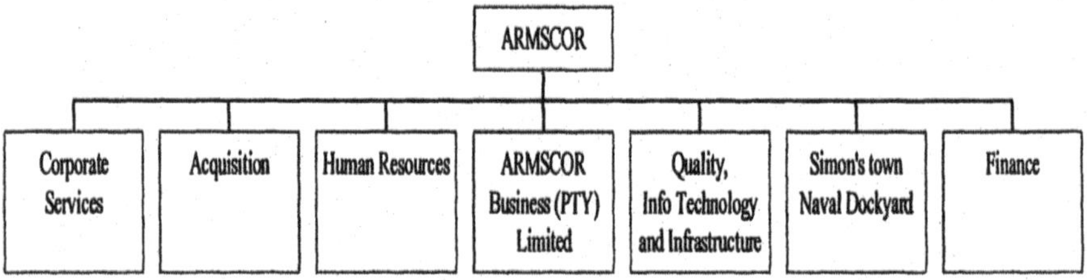

Figure 4.2. Armscor Corporate Structure in 2004

Unlike its function in the *apartheid* years, Armscor no longer had a significant production role (having given that up to Denel in 1992) though its subsidiaries still produced small amounts of highly specialized materiel.[9] Rather, beyond its basic function of materiel acquisition, maintenance and disposal, Armscor's outreach to the larger defense community now was mainly concerned with research, development, testing, evaluation, and technical consulting. It maintained world-class facilities and technical staff for these roles, which were played largely by Armscor Business (Pty), set up in 2002 to manage what the organization calls its "strategic facilities." [10] By 2004, Armscor Business consisted of fourteen subsidiaries offering a broad range of reimbursable services to the defense industry at large. (See Figure A.1. for the structure of the Business grouping, and Appendix A for more detailed descriptions of the individual subsidiaries.)

Despite the arms export scandals of the mid 1990s, and the loss of the prerogative to oversee arms exports, Armscor continued to play a role in marketing South African defense technology. This role included the facilitation of arms exhibitions such as the biennial Africa Aerospace and Defence show (billed as the largest exhibition of military materiel on the African continent) which Armscor hosted in 2002 and 2004.[11] Armscor's officials also participated in conferences on exchanges of military technology between South African and other countries, such as those facilitated by the bilateral US-South African Defense Committee (DEFCOM) or the Trilateral Commission of the Ibsa Dialogue Forum between South Africa, Brazil and India.

The separation of Armscor and Denel in 1992 was accomplished by executive fiat, leaving a number of legal ambiguities and unanswered questions that haunted Armscor's role for over a decade. Armscor's activities during that period were complicated by a lack of legislative authorization and clear mandate for its new functions. The founding legislation in 1968 had specified its role in detail, including severe limits on its prerogatives to compete with the private sector in offering goods and services. The arrangement that included the activation of Denel in 1992 restricted Armscor from defense production altogether.[12] But the old legislation of 1968 and the executive fiat of 1992 did not adequately capture the realities of Armscor's new functions in the post *apartheid* era. By the early years of the 21st Century, it was clear that Armscor's long-term viability could be much enhanced by its ability to offer services to clients inside and outside of the public sector. After some years of confusing limbo, Armscor obtained a new mandate in 2004, with legislation that went far in resolving the various legal ambiguities.[13] However, fully defining and implementing the new vision was likely to require some years.

In 2005 Armscor was pursuing several promising commercial possibilities. It offered a range of services to clients within South Africa's public and private sectors. It also allowed other countries to use its research, test and evaluation facilities, in effect renting Armscor human and infrastructural resources to carefully vetted foreign clients. Another Armscor initiative involved efforts to offer materiel acquisition expertise to other African countries. The opportunity to sell this service surfaced in multilateral agreements within both the Southern African Development Community and the African Union. Armscor's efforts to market its expertise to these other potential African consumers appeared to be in the earliest stages in 2005.[14]

Armscor's status as an embedded component of the Department of Defence has come at a cost. Most of the organization's expenses in 2005 still were funded by Defense allocations. Senior managers had argued for years that Armscor suffered from insufficient funding.[15] Meanwhile, its responsibilities had steadily increased, particularly in the oversight of the Defence Packages. As the military services themselves faced funding shortfalls for their activities, the Department of Defence turned increasingly to Armscor for assistance in maintaining or disposing of Defence Force materiel. The increase in tasking had not been accompanied by increases in resources until about 2003. According to industry insiders, one result had been a severe limit on Armscor's ability to recruit and retain needed talent.[16] Another result had been a resort to the private sector to perform some of the maintenance and systems engineering tasks for the Department of Defence that Armscor had been chartered to perform.[17]

The senior leadership of the Department of Defence apparently acknowledged Armscor's financial straits, augmenting its funding by almost 20 percent over the period 2003-2004, and indicating that the funding would be augmented another 12 percent in each of the subsequent two years.[18] In 2005 Armscor's managers believed the institution's problems were temporary. They expected that shortfalls would be at least partially solved by increases in Departmental funding and by the commercial options authorized in the 2004 legislation.[19] According to an industry insider, in 2005 the Department of Defence was working on a new Defence white paper intended to clarify and better define the Armscor role.[20] Even so, the jury was still out on Armscor's long-term future.

Denel

A second major public sector entity in South Africa's defence industry is Denel, founded in 1992 when the government separated Armscor's production subsidiaries from its other activities.[21] Unlike Armscor, which remained a subsidiary of the Department of Defence, Denel was moved to the Department of Public Enterprises. In 2005 it still was a state-owned company, with the state as the sole shareholder, although it was viewed in South Africa as likely to undergo some process of privatization in the future.[22] A thirteen-person board of directors met at least once a quarter to oversee corporate policy and strategy for Denel.[23] Day-to-day management was vested in ten-member executive committee.[24]

With some eighteen subordinate enterprises in 2005 (twelve directly involved in defence production or testing), Denel was by far the largest corporate actor within South Africa's defense-related industries.[25] It also played a central role as a contractor and subcontractor within that community. Denel's subsidiaries were scattered across the country, though a majority were located in Gauteng Province, around Tshwane (Pretoria) and Johannesburg.[26] The corporation produced a large spectrum of air, ground and naval combat equipment, ranging from weapons systems to avionics to high technology command and control to electro-optical and laser applications. Some of its products have proven very competitive in world arms markets, particularly its artillery systems. (See Appendix A for additional detail.) Although Denel's subsidiaries retained their own sophisticated production facilities, they also depended heavily on subcontracting from private sector producers. And Denel's subsidiaries produced subassemblies on contract for other defense industry consumers, continuing the private/public sector linkages that evolved in the *apartheid* years.[27]

Denel underwent considerable restructuring and change in the years after its establishment in 1992, shedding some former Armscor enterprises, closing down others altogether and consolidating still others.[28] These efforts reached a crescendo in about 1999 and continued into 2004, when sweeping changes were made in the corporate structure in a transparent effort to improve profits. Since its founding, Denel also has been subjected to a continual process of personnel reduction. It employed some 16,400 people in 1992, but by early 2005 had reduced that total to about 9,300.[29] Like most of the rest of South African industry, Denel has been engaged in a variety of "affirmative action" programs to provide opportunity to employees from historically disadvantaged communities. (In its 2004 *Annual Report*, Denel reported that its workforce was 48.4 percent white and 51.6 percent black, claiming an increase in "black management" from 35 to 65 individuals over the previous year.[30])

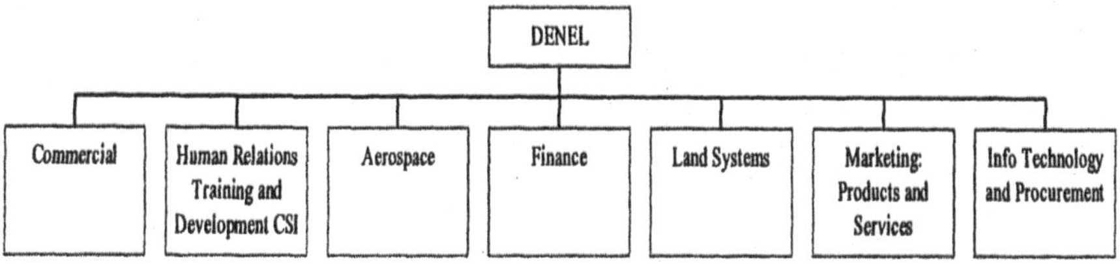

Figure 4.3. Denel Organizational Structure (As of April 2004)

Because of Denel's large size and centrality to the industry as a whole, the seemingly endless changes have had significant reverberations within South Africa's arms producing community. According to arms industry insiders, this volatility has generated considerable angst within Denel itself and among its numerous subcontractors and partners in the private sector that depend on Denel contracts. The turbulence in the organization has contributed to a steady hemorrhage of mid-level engineers and technicians to the private sector, seeking better remuneration and less employment uncertainty.[31]

In its quest for profits, Denel has vigorously pursued export opportunities. Those efforts have seen growing success. In its 2003 *Annual Report*, Denel claimed an increase in export sales of almost 20 percent over the previous year, also noting that export sales had risen to 56 percent of total sales, a four percent increase in one year. By 2004 sales had risen to 58 percent of total sales.[32] Related to its emphasis on export marketing, Denel has sought out relationships with various external defense firms. (In 2002, it claimed "strategic" partnerships with thirteen foreign arms enterprises, although in subsequent years it was less specific. The 2004 *Annual Report* simply mentions representative partnerships with the US-based General Dynamics, the German-based Diehl Munitions and the UK's Radway Green.[33]) Some of Denel's foreign partnerships are a direct result of the South African "Strategic Packages" arms acquisitions, and South Africa's foreign weapons purchases have benefited Denel directly and profoundly (see Chapter 3 and Appendix A for additional details). But Denel also has penetrated several niche markets purely on the merits of its technology. This includes the partnership with the US-based General Dynamics Ordnance and Tactical Systems (GD-OTS) to produce ammunition and innovative new artillery systems, including main gun technology tested in the United States in 2004 for the US Army's new *Stryker* armored vehicle.[34]

Denel's status as a public sector enterprise has benefits and costs. While it enjoys more security from market pressures than a typical civilian commercial firm accountable to impatient shareholders, it cannot make the rapid, socially disruptive economic decisions demanded by a volatile market. Partly as a result, it has struggled to achieve corporate profitability.[35] It showed a net loss before taxes in fiscal year 2003 of just under US$6 million (compared to the previous year's net loss of US$44 million) but the loss in 2004 had escalated once again, to US$61 million.[36] Still, a decade of restructuring has resulted in a much more market (and export) oriented organizational culture, and the organization's exports have risen steadily. Gross revenues in 2003 were up 12 percent from the previous year, and by 2004 had increased another 6.7 percent.[37] Botha concurs with Denel's management that the corporation's long-term economic prospects are quite good.[38]

Defencetek

A third public sector entity amongst the defense related industries is Defencetek, a subsidiary of the Council for Scientific and Industrial Research (CSIR). The parent organization was founded in 1945 with a subsequent long relationship to South Africa's military.[39] However, Defencetek itself is a relatively new subsidiary – CSIR created it in 1999 as the subordinate organization that would capture the CSIR interests in defense and security. Defencetek is headquartered in the large CSIR complex in the Tshwane (Pretoria) area.

At the time of its founding in 1999, Defencetek was designated by the Department of Defense as a Defence Evaluation Research Institute (DERI), bestowing on it an official role in maintaining the "strategic capabilities" articulated in the *Defence Review* and *White Paper on South African Defence Related Industries*. In fact, Defencetek claims in its literature to be the host of two DERIs: an Electronic Evaluation and Research Institute (EDERI) and a Landwards Evaluation and Research Institute (LDERI).[40]

According to its director in 2003, Defencetek views itself as a consultative body on issues of sophisticated technology. In that role, it uses its expertise and facilities to provide "smart buyer" and "smart seller" advice to clients, particularly within the South African public sector – and including military and police. It also offers advice on counter-trade issues, both to the South African government and to foreign and domestic clients in the private sector. CSIR traditionally has emphasized "pure" science, research and consulting services rather than materiel production, and Defencetek operates within that tradition. Other than prototypes, Defencetek did not market materiel *per se* in 2005. However, it invested in its own intellectual property. It conducted research on security applications, maintained the capabilities for independent testing of high technology defense materiel, and studied the doctrinal implications of new security technology.[41]

Among other initiatives, the organization's analysts had concluded that the future will see a strong demand for protection against landmine threats, and the organization had devoted substantial research to vehicular and personnel protection from landmines. Its interests extended as well to landmine clearing technology. These are areas in which South Africa already possesses "cutting edge" technological advantage, resident in other enterprises such as Denel's Mechem and BAE Land Systems OMC, but Defencetek has developed commercially attractive niche applications, including fuel-air devices for mine clearing and minefield breaching, explosive device disrupters and simulants for mine protection evaluation. Among its clients for countermine technology in 2005 were the US Department of Defense.[42] (For the relevant details, see Chapter 3.)

Defencek's interests in 2005 were very broad. It had advised the South African military on combat clothing, including camouflage patterning, attenuation of infra-red detection and anti-camouflage combat identification, and it had studied the feasibility of locally produced non-lethal weapons. The enterprise emphasized its expertise in such high-tech areas as aerodynamic research, information technology, communications/electronics and countermeasure applications.[43]

Defencetek has vigorously pursued its relationship with the South African military, billing itself as "independent, impartial technical consultants" to the Department of Defense, and casting itself in the role of technical adjuncts to the Department. According to its director, formalization of this relationship still was in process in mid 2003. Since Defencetek provides advice on security-related technology without belonging to the military establishment, its leadership believes it offers objectivity unavailable from organizations (like Armscor) within the Department of Defence itself.[44]

Defencetek in 2004 and 2005 was still in somewhat of a transitional period. It apparently could no longer depend on government funding to fully meet its needs, and was beginning an aggressive effort to market its services to domestic and foreign clients. Defencetek itself was something of an unintended metaphor for South Africa's armaments industry. It was not part of the Department of Defence, but was closely related to it. It was not fully "public sector," yet not exactly "private sector." At least in theory, it existed primarily to service South Africa's need to maintain strategic capabilities for the national military establishment, yet it depended on private sector (and even foreign) clients for much of its income. In its role as a DERI (or as a host of multiple DERIs) it played a part that had been carefully scripted in South Africa's policy documents. Yet it was feeling its own way in the new South Africa.

Major Defense Related Industries in the Private Sector

The public sector defense industry entities – Armscor, Denel and Defencetek – are a complex reality. But they are simple compared to the private sector. Here, the arms industry since 1994 had been a diverse and constantly shifting universe of corporate consolidation, divestment, renaming and restructuring. In the past, private sector enterprises offering security-related materiel and services typically were subsidiaries of large, highly diversified corporate groups. Today's private sector defense industry still includes corporate groups of defense-related companies, although it also contains a variety of small, independent enterprises.[45]

The larger private sector "defense" actors include African Defence Systems (ADS), Advanced Technologies and Engineering (ATE), Aerosud, BAE Land Systems OMC, Grintek and Reunert. No treatment of South African armaments producers would be complete without some assessment of such prominent industrial actors. (For additional detail on several of these organizations, see Appendix B.)

African Defence Systems (ADS)

In 2005, African Defence Systems maintained its corporate headquarters in Halfway House (between Tshwane and Johannesburg). Previously known as Altech Defence Systems, the larger Altech group had divested itself of its defence subsidiaries, and these ultimately were purchased by foreign investors. ADS was now a member of the Thales Group International (formerly Thomson CSF), a French-based multi-national corporation.[46] In its earlier forms within Altech, this enterprise had a long association with the South African defense establishment as a private sector producer of electronic and cyber applications.

ADS specialized in electronic, simulator and weapons command-and-control equipment. Among its products were a state-of-the-art artillery fire control system and training simulators for weapons, aircraft and vehicles.[47] Its specialties included artillery command and fire control (with its flagship AS2000, a state-of-the-art, computerized system typically offered for sale with Denel's sophisticated artillery), air defense and naval command and control systems and weapons integration. The firm produced a range of sophisticated simulators for crew training in aircraft, armored vehicle, naval and air defense systems. Its services included modernization and upgrading of weapons and command and control systems manufactured elsewhere.[48] This South African-based enterprise can be assumed to have access to state-of-the-art French technology in its fields of interest.

Advanced Technologies and Engineering (ATE)

Another significant South African actor in the avionics field in 2005 was ATE – Advanced Technologies and Engineering Company (Pty) Ltd. – a firm with long ties to the South African Air Force. In 2004, ATE maintained its South African headquarters in Halfway House, between Johannesburg and Tshwane (Pretoria). The firm described itself as "the South African component of a private-sector systems engineering and management group," claiming expertise in designing and integrating advanced systems for various combat aircraft, remotely piloted vehicles and armored vehicles.[49]

ATE was a world leader in systems integration of advanced components and weapons for aircraft (including remotely piloted vehicles) and ground combat vehicles.[50] It had been awarded substantial contracts in South Africa and abroad for upgrades of first-line combat aircraft, which it equipped with new avionics and weapons. Among ATE's other offerings were a sophisticated remotely piloted vehicle (the *Vulture*) designed primarily for artillery targeting and used by the South African Army with its G-5 and G-6 artillery weapons. ATE also built an anti-tank guided missile turret for armored vehicles, and by 2005 had won contracts for upgrading the Soviet-made T-72 main battle tanks and BMP infantry fighting vehicles of at least one other African nation.[51] But its most commercially successful programs appeared to be avionics and weapons system upgrading. In 2001, ATE won a major contract to upgrade the Mirage F1 multrole fighters in service with the Spanish Air Force.[52] It also offered tailored upgrades to the Russian-designed Mi-8, Mi-17, Mi-24 and Mi-35 helicopters.[53] According to industry insiders, Russian aviation industry officials had complained about the loss of their opportunity to provide such upgrades, since ATE apparently offered superior quality and options to prospective customers.[54]

At the beginning of the 21st Century ATE was partially foreign-owned, BAE having purchased a 20 percent share. However the South Africans dissolved this partnership in 2003 when both enterprises found themselves in competition for lucrative contracts to upgrade Eastern European military aircraft, and ATE purchased the BAE shares.[55] The South Africans believed their equipment, service and terms were superior. Still, the earlier BAE connection probably accounted for the fact that ATE was a major beneficiary of the Defence Packages, having won the subcontract (worth some $80 million) to design the South African mission package for the BAE-furnished *Hawk* lead-in fighter-trainer aircraft.[56] (ATE also furnished the advanced avionics for Denel's *Rooivalk* helicopter and Pilatus PC7 trainer in service with the South African Air Force.)

ATE's commercial activity offered a particularly good example of a South African niche capability – the willingness and ability to provide very sophisticated upgrades to older combat systems, resulting in substantially enhanced combat effectiveness. As evident in ATE's marketing, the South Africans appeared willing to do this on a scale smaller than their potential competitors in the developed West. They also are able to subcontract quickly and effectively across a range of different industries to "tailor" packages for specific client requirements.

Aerosud

In 2005, Aerosud still was a fairly young but very successful aeronautical engineering company, formed by designers, engineers and technicians (largely from Armscor) that had been involved in South Africa's most ambitious military aviation project of the time – the *Rooivalk* combat helicopter. The company subsequently served as the lead agency in a project sponsored by the South African Air Force to adapt the Russian Klimov RD 33 jet engine (used in the Mig-29 air superiority fighter) to South Africa's Mirage F-1 and Mirage 3/*Cheetah* combat aircraft.[57]

Aerosud maintained a focus on military aircraft engineering but had expanded into civil aeronautical applications, including design, development, prototyping, manufacture and support. It specialized in upgrades, life extension programs and airframe/avionic integration. In 2001, it won a major Boeing contract to manufacture parts for Boeing's commercial aircraft, and in 2003 it won an equally significant contract to manufacture subassemblies for the enormously successful European-produced Airbus A320 commercial jets. In late 2004, Aerosud Managing Director Paul Potgeiter informed the local South African media that his firm had committed itself to "significant" agreements with the (European) Airbus Military firm for design and manufacture of components for the future A 400M military transport.[58] Aerosud's facilities were located near Waterkloof Airbase in Centurion, just outside of Tshwane (Pretoria).

BAE Land Systems OMC

In 2005, South Africa's major corporate actor in the production and refurbishment of security vehicles was BAE Land Systems OMC. The South African parent of this enterprise was Land Systems South Africa, a business unit of the British-based BAE Systems Land Systems conglomerate.[59] As this is written in early 2005, BAE Systems owns a 75 percent share of the South African parent, while a local investment group holds the remaining 25 percent.[60]

Land Systems South Africa is managed by South Africans and enjoys substantial autonomy, making its own management, production and marketing decisions, while the larger British connection gives it overseas market access.[61] It consists of two subordinate enterprises, Land Systems Gear Ratio and Land Systems OMC. OMC is the enterprise that largely has consolidated the separate South African enterprises in producing military vehicles.[62]

Land Systems OMC is an acknowledged world leader in mine-protected applications and has pioneered in the study of mine effects on vehicles. It also performs much of the vehicle refurbishment and upgrading for the South African Army and Police and has seen substantial success in marketing its security vehicles abroad, having penetrated even the notoriously difficult North American market in partnership with General Dynamics Land Systems-Canada.[63] Its most recent success, described in Chapter 3, was the sale of 148 mine-protected vehicles to the US Army in 2005, with the prospect for substantial additional sales. OMC maintains its headquarters and production facilities in Benoni, a few miles east of Johannesburg. (For additional detail and background this firm, see appendix B.)

Grintek

Grintek Limited is an electro-technology conglomerate with deep roots in defense production in South Africa.[64] In 2005 its corporate headquarters were in Centurion, in the Tshwane area. Grintek's interests ranged across a spectrum of private sector telecommunications and electrical power applications. Several of its daughter enterprises were heavily engaged in production of military materiel, building state-of-the-art signals intelligence, electronic warfare, communications and other electro-optical equipment. It was an acknowledged world leader in the miniaturization of aircraft communications.

Grintek had been a direct beneficiary of the South African Defence Packages, having been selected to provide significant amounts of materiel for installation in each of the different "Package" systems. In addition to the contracts themselves, "Packages" partnerships and the related foreign ownership had furnished substantial access to foreign markets that had vastly expanded the firm's export opportunities. Grintek's high-technology equipment was sold around the world in European-produced military aircraft and ships.

In 2005, Grintek and its subsidiaries also were substantially foreign-owned, the Swedish industrial giant Saab owning some 22 percent of Grintek Limited and 49 percent of its daughter enterprise Grinton.[65] (In February 2005, Saab offered to buy Grintek outright. If accepted, this offer would give Saab a control ling 70.3 percent share of the South African group.[66]) In 2005, the German firm Ewation owned 45 percent of another Grintek daughter enterprise – Grinton Ewation. (For additional background and detail on Grintek, see Appendix B.)

Reunert

Reunert Ltd is another large South African conglomerate with a long history of military materiel production in a variety of fields.[67] In 2005 only four of its eighteen daughter enterprises (and seven hundred of its five thousand employees) were substantively involved in arms production, mainly those grouped in the Reutech subdivision. Reunert/Reutech remained a large-scale producer of sophisticated fuses for artillery, rockets and guided missiles. Its subsidiaries also built search and tracking radars and offered logistics support for artillery and fire control systems.

Like most of the other major arms producers, Reunert had been a significant beneficiary of the Defence Packages: one of its daughter companies had been contracted to furnish the tracking and fire control equipment for the corvettes. That subsidiary – Reutech Radar Systems (RRS) – was partially owned by the

European aerospace multinational EADS, giving this South African enterprise a significant window to the world market for military radar and tracking systems.

Perhaps not surprisingly, in 2005 most of Reunert's armament production was exported, and the firm had seen considerable and growing worldwide interest in its military materiel. Reunert's stabilized platforms for optronic and radar trackers were standard features on German-produced K130 class corvettes. EADS also incorporated the Reunert stabilized platform and antenna housing in the radar it had been contracted to install on a prototype of the US Navy's new Littoral Combat Ship (LCS).[68]

Reneurt's corporate headquarters were in Sandton (Johannesburg) and its daughter companies were widely dispersed across the country. (For additional background and detail on Reunert's defense-related subsidiaries, see Appendix B.)

Other Defense-related Industry Actors

In addition to the larger arms producers noted above, many small South African firms in 2005 offered goods or services with military applications.[69] Botha characterized these as "skills-based, entrepreneurial companies that [lived] off contracts from the majors."[70] A large number of the smaller defense related enterprises were direct beneficiaries of the Defence Packages in their own right or indirect beneficiaries as sub-subcontractors for larger enterprises like Denel. Botha called attention to a substantial decline since 1993 in their overall number, but the quantity of enterprises engaged in some form of security research and production still was remarkable for a developing nation. Botha's research indicated that the successful firms within this category tended to be very export-oriented.[71] A comprehensive listing of the smaller defense related firms would be tedious; but it is possible to get a sense of their diversity and contribution by noting some significant product offerings.

One evident aspect of the South African defense related industry, including particularly the smaller enterprises, has been the heavy representation of cyber, simulator and high-end communications-electronics applications. This is exemplified by enterprises like C^2I^2 Systems of Rondebosch that specialize in software for combat information and management systems.[72] C^2I^2 has successfully marketed its software to NATO consumers like the US-based Raytheon Corporation. Similar enterprises include Sysdel Close Corporation, headquartered in Centurion, specializing in the development of electronic warfare systems and software. Sysdel, for its part, had by 2005 developed significant partnerships with major European and US defense contractors.[73]

Other South African producers of military materiel spanned a considerable range of different applications. Some offered high-quality avionics equipment, components, and aircraft upgrading options. These included Aerospace Monitoring Systems (AMS), located in Halfway House between Tshwane and Johannesburg.[74] AMS had successfully marketed its aircraft diagnostic equipment to

NATO consumers and Australia. South Africa's defense-related production included exotic materials like advanced polyurethane wheel technology for combat vehicles, specialized lubricants, and special-purpose boats. Small companies also offered a rich inventory of conventional products ranging from military clothing to weapons systems of various kinds. (See Appendix C for a selective listing of actors and some additional detail.)

Among South African enterprises in 2005 were businesses that the South Africans themselves did not seem to consider defense-related, but which had clear potential for military applications. These included the air transport company Safair, at one time the owner of the world's largest fleet of US-produced L-100 (Hercules) aircraft – the civilian version of the military C-130.[75] During the *apartheid* era, Safair had a much closer (if covert) relationship with the South African military. As a "civilian" enterprise, its aircraft could legally fly over parts of Africa that were "off limits" to the South African Air Force, a capability the South Africans used for airlift of personnel and supplies to their forces in Southwest Africa (now Namibia).[76]

The "Benchmark" Products

Any discussion of South Africa's arms industry – and its capabilities – would be incomplete without acknowledging several of its most notable achievements, successes categorized here as "benchmark" products.[77] These provide useful, historically based insights into South Africa's capacity to conceive, design, and produce military technology, and also suggest some limits to that capability. Any list is inherently subjective, of course, and the items offered here might be as much an indication of the author's interest as of South Africa's capabilities. Yet the benchmark products provide some basis for making comparisons to the arms industries of other developing countries. A number of different South Africans programs and products are good candidates for the list, but the nuclear weapons program, the G-5/G-6 gun system, the *Rooivalk* helicopter and the SMR 95 jet engine certainly represent strong contenders. They are the ones cited here.

The Nuclear Weapons Program

South Africa holds several unique distinctions in regard to nuclear weapons. At the beginning of the 21st Century, it was the only African country that had developed such arms. It also was the only country in the world that had produced deployable nuclear weapons only to discontinue the program and dismantle all the weapons under international supervision.

Nuclear weapons were first produced in South Africa in the early 1980s and dismantled in 1991, the year that South Africa acceded to the nuclear Non-

Proliferation Treaty.[78] A variety of explanations have been offered for the decision to terminate the program, including the cynical suggestion that leaders of *apartheid* South Africa did not want to endow their political successors with nuclear weaponry. A more credible explanation would cite the considerable pressure on the de Klerk administration by the governments of the United States and (probably) Israel to prevent nuclear weapons from falling into the hands of an ANC government that might be inclined to provide them to radical states like Libya.[79] However, individuals closely connected to the program at the time tend to describe the chief motivation for denuclearization as the government's determination to downsize the defense budget. They hardly need to mention the international demands on South Africa to sign the Non-Proliferation Treaty. Nor is it any deep secret that by 1990, the threat that motivated the decision to initiate the program had ceased to exist.[80]

When State President F. W. de Klerk made the decision to terminate the nuclear weapons program, South Africa's nuclear inventory contained six deliverable weapons, all of them low-yield (15-20 kiloton) fission devices, the first more-or-less deployable one having been built in 1982.[81] Had South Africa actually used one of these weapons, it almost certainly would have been delivered by air, probably mounted in a television-directed glide bomb dropped from one of the South African Air Force's aging British-made *Buccaneer* bombers. However, had South Africa not discontinued its space program, it probably could have delivered its nuclear weapons by intermediate-range ballistic missile before the end of the 1990s.[82] (The de Klerk government discontinued the space program in 1993.[83])

South Africa's acquisition of nuclear weapons and the dismantling of those weapons has been cited by scholars as a highly significant case study.[84] That broader theoretical discourse, however, is outside the direct interest here and will be left to others to dissect and debate. Rather, the South African nuclear weapons program is one particularly illustrative indication of the country's ability to conceive, design and create military armament, and the book turns to the details of that story.

At the beginning of the "nuclear age," South Africa found itself with plentiful supplies of uranium, a sophisticated industrial economy and a gifted community of scientists and engineers eager to take advantage of emerging new technology.[85] A substantial internal constituency for the development of a nuclear industry existed in the country by the 1950s, including prominent members of the civil government, military and scientific and business communities.[86] Yet the country's interest in nuclear weapons grew out of several later convergent chains of events, and early interests were more explicitly in uranium mining, research and peaceful applications of nuclear power than weapons *per se*. But the earliest South African government initiatives clearly framed nuclear energy as a strategic resource to be carefully controlled.

Upon its accession to power in 1948, the new National Party government had moved quickly to consolidate government control over nuclear issues. A foundational Atomic Energy Act (1948) granted the government a monopoly on control of nuclear materiel and information, even to the point of authorizing secret trials for those who violated laws against disclosure of such information. The Act also established a national Atomic Energy Board (later recast as the Atomic Energy Corporation) to oversee the state's nuclear equities.[87]

By the early 1950s South Africa had sent a large number of scientists and technicians to the United States to study nuclear technology, beginning a cooperative relationship on peaceful uses of nuclear energy that lasted until 1976. Other scientists and students went to the United Kingdom, establishing cordial ties between the communities of nuclear scientists in the three countries that lasted well into *apartheid* era. South Africa's nuclear scientists also ultimately developed relations with French and German counterparts as the nuclear industries in those countries slowly coalesced in the 1950s and 1960s.[88]

In 1952 South Africa opened its first uranium production plant (predictably, with assistance from the United States and United Kingdom) and by 1955 had built an additional fifteen plants to produce uranium, which had by that time become an important export commodity. In 1957, South Africa and the United States formalized their nuclear cooperation in an agreement that included US supply of a research reactor.[89]

In 1959, the South African Cabinet authorized the national Atomic Energy Board to commence the country's own research and production program, a measure that ultimately resulted in construction of a second research reactor at Pelindaba, west of Pretoria (completed in 1965) and later, construction of two nuclear power plants in collaboration with French contractors at Koeberg near Cape Town (completed in the 1980s). By 1969 South Africa also was building a highly secretive uranium enrichment plant at Valindaba, adjacent to its Pelindaba facility, an activity first identified publicly in 1970, but not fully operational until the end of the 1970s.[90]

The timing of South African decisions to commence substantive research and development of nuclear weapons has never been explained to the outside world. It seems eminently reasonable to assume that some weapons-related research had commenced by the early 1960s. Purkitt and Burgess seem to associate the political decision to invest in nuclear weapons with the construction of the Valindaba enrichment plant in the late 1960s.[91] South Africa's Atomic Energy Board (AEB) supervised the country's nuclear activities during this period, and until 1978, the Department of Defence (and, thus, Armscor) had no direct responsibility. However, the AEB worked closely with Armscor subsidiaries such as Kentron and Somchem in the early 1970s to develop a gun-type initiator that ultimately was successfully tested in 1974.[92]

The South Africans clearly were contemplating the acquisition of nuclear weapons by the early 1970s. There are unverified claims of the existence of South African nuclear weapons in the 1960s, presumably in collusion with Israel. However, the decision to "weaponize" seems more likely to have been taken by Prime Minister John Vorster in 1974, when it became clear to South Africa's nuclear scientists and senior policymakers that local production of nuclear weapons was feasible.[93]

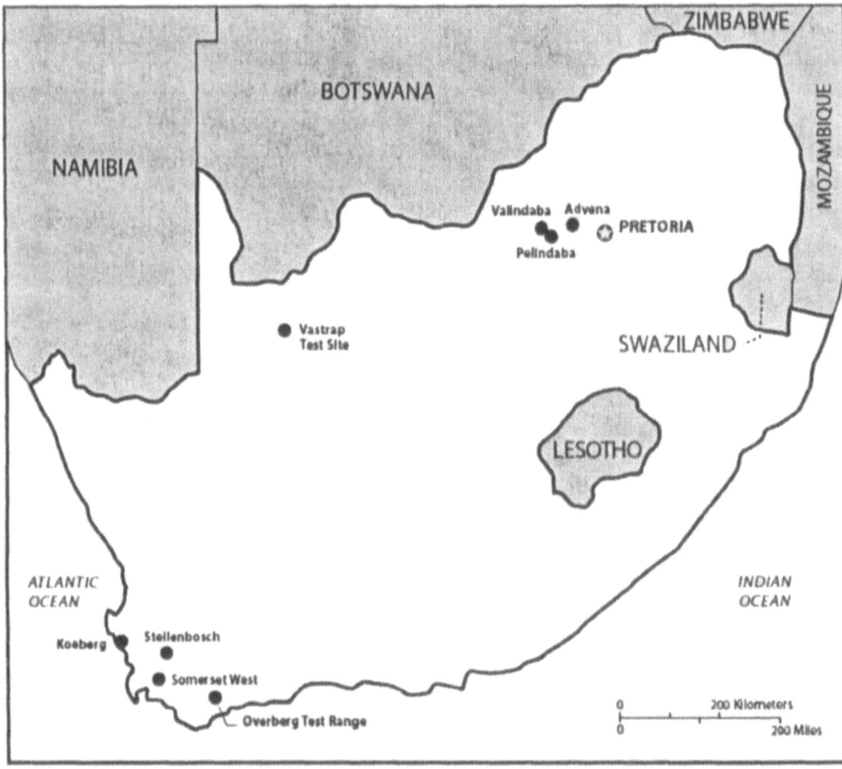

Figure 4.4. Key South African Nuclear-related Sites (1960-1990)

By 1974 the Atomic Energy Board already had constructed nuclear test devices, presumably for peaceful applications of nuclear power. By the mid 1970s the Board had selected and prepared an underground nuclear test site at the remote Vastrap military base in the Kalahari desert area in the northern Cape Province. In 1977, the South Africans attempted a secretive dry run of an underground nuclear test. The test preparations were detected by Soviet satellite surveillance, and in a show of unity rare for the Cold War era, the United States, the Soviet Union and various other countries successfully pressured South Africa to discontinue the test.[94]

The unhappy results of its intervention in Angola in 1975 and 1976 turned South Africa's attention more explicitly to strategic implications of nuclear weapons.[95] The country's political isolation in the mid 1970s coincided with increasing international consensus on the merits of non-proliferation, led by the members of the existing "nuclear" club, particularly the United States. In 1975 the United Kingdom terminated a significant program of naval cooperation. By

1978, the US had ended its connection on issues of nuclear energy due to South Africa's refusal to sign the nuclear Non-Proliferation Treaty. A presidential election in the United States in 1976 had brought in a new administration (that of Jimmy Carter) much less inclined to cooperate with South Africa on any issue and much more inclined to criticize its policies. The timing of these events, along with the growing East Bloc presence in Angola, reinforced South African perceptions of abandonment by the West, and seems to have boosted the nuclear weapons program. The secretive South African nuclear programs now became even more secretive and much more weapons-oriented.

The country's interest in a nuclear deterrent was enhanced by a key leadership change in 1978. In that year, P.W. Botha, who had just overseen the vast expansion of South Africa's arms industry in his previous twelve years as Minister of Defence, assumed the position of Prime Minister. Almost immediately after his elevation, Botha established a cabinet committee to oversee a nuclear weapons program, with representation from the Defence Force, the Atomic Energy Board and Armscor. Various working groups also were created to address different aspects of a weaponization program – policy issues, strategy issues and technical issues. In his tenure as Defence Minister, Botha had supervised the creation of South Africa's arms industry under the management of Armscor, so not surprisingly, this organization now began to play the key role in nuclear weapons production. One immediate result was Armscor's authorization to construct a new facility some 20 kilometers west of Pretoria (first called Kentron Circle and ultimately named Advena) commissioned in 1981. This would be the site for constructing nuclear weapons and testing diagnostics.[96]

Although South Africa has never admitted testing any nuclear devices, a US surveillance satellite detected a highly unusual "double flash" anomaly over the South Atlantic in September 1979. This event may have been a South African test (or an Israeli test with South African collusion). If it were a nuclear test, it infers a more sophisticated nuclear technology than South Africa is known to have had at the time, although the country's capabilities were routinely underestimated by outsiders. The anomaly has never been adequately explained.[97]

South Africa's nuclear weapons program was veiled in intense secrecy and highly compartmented; its details were confined to a small inner group of military officers and civilian technicians. The team that actually built the nuclear weapons at Advena at first numbered only about 35, growing slowly to about 100 by 1989. The small size of the team, the limited amount of highly enriched uranium (HEU) produced in South Africa and the technical difficulties of developing an extraordinarily complex program with minimal external assistance imposed severe limits on the rate of production. When the program was terminated in 1990, Armscor had turned out a total of eight low-yield devices classifiable as "weapons." Five were deliverable nuclear arms, another was a test device, and still another was under construction as a test device. A final weapon had been constructed for training purposes, but lacked an assigned HEU core.[98]

While the details of the nuclear weapons program itself were closely held state secrets, the South Africans were concerned from the beginning about the options for delivering their weapons. This concern provided a strong impetus to develop two delivery systems intended to complement the nuclear capability, but possessing considerable military utility in their own right and thus offering some plausible deniability. One system was a ballistic missile (based on the Israeli *Jericho* II design and developed with Israeli assistance) purportedly to provide a low-earth orbit reconnaissance satellite capability. The other was a television-guided glide bomb under development by the South African Air Force. The South Africans seem also to have briefly flirted with the prospect of an artillery-delivered nuclear capability, though this option apparently never progressed beyond conceptual study. Nonetheless, the missile and "smart bomb" programs energized a wide circle of Armscor subsidiaries and contractors, and resulted in significant new technology.

Another Armscor subsidiary – Houwteq – was created to develop and apply the missile technology, with facilities near Houw Hoek and Stellenbosch in the Western Cape region. As part of this effort, Armscor also developed the Overberg Test Range along with a test flight and development center near Bredasdorp (collocated with an Air Force test site) overlooking the Indian Ocean. Houwteq would not survive the political transition of the 1990s, but the Overberg Test Range and test flight center were still operational in 2005, and in recent years, the South Africans have endeavored to market this site and its associated technical infrastructure to foreign clients. (In fact, by early 2005 there were indications that Denel planned to revive the Houwteq enterprise as a center for development of a communications/earth resources satellite program.)[99]

The Armscor subsidiary Somchem was tasked to develop the ballistic missile propulsion. It expanded its facilities at Somerset West (also in the southern Cape area) to build the rocket motors. It also built a test facility in the nearby Hangklip Mountains where it could test them.[100] These facilities survived the termination of the missile program in 1993, and Somchem still produces missile power plants and propellants.

Kentron, the Armscor subsidiary with long-standing expertise in missile guidance technology, was involved in the research and development programs for both the ballistic missile and "smart" bomb. Its investment in inertial navigation systems and other missile technology had continuing application well beyond the delivery of nuclear weapons, and Kentron's descendant enterprise in Denel still builds the state-of-the-art tactical missiles. It probably is fair to say that the emphasis on the nuclear delivery systems contributed significantly to its current expertise and capability.[101] The Armscor subsidiary Eloptro seems to have been involved in work of the associated reconnaissance satellite capability and presumably other electro-optical applications.[102]

South Africa's nuclear weapons program continued for about twelve years and ended after South Africa initiated substantial political reform. During this period, Armscor played the key role in producing the weapons and determining the kind of nuclear capability that South Africa would have.

The involvement of Israel in South Africa's defense relationships has been a subject of an interesting (and at times overheated) literature with considerable conjecture and much less irrefutable fact.[103] This particularly is true of the nuclear weapons program. The Israelis clearly were involved in the ballistic missile program at various levels. Almost certainly they contributed in a variety of other ways to the development of nuclear weapons in South Africa, whether by exchanges of information, technology and materiel or by more tangible relationships, or perhaps by encouraging the development of the weapons themselves. However, there is little evidence that the Israelis – or any other foreign actors – were fundamentally a part of Armscor's processes of nuclear weapons production. It is important to emphasize that the nuclear weapons program is a good benchmark precisely because it was a South African program with South African antecedents, South African expertise and South African technology. In these respects, it was very unlike (for instance) the nuclear weapons produced by Pakistan, described by one scholar as "either stolen from a European nuclear facility or provided by China."[104]

South Africa's nuclear weapons program has been criticized for lack of imagination, and it is true that the sophistication of the weapons applications never advanced significantly beyond the nuclear technology pioneered by the South African Atomic Energy Board in the early 1970s. The program also has been characterized by some of its participants as haphazard, poorly managed and inefficient.[105] Whatever the merits of these criticisms, they profoundly miss the point. South Africa's leaders were seeking a credible nuclear capability that could be demonstrated if required, yet a capability sufficiently small-scale and ambiguous that it would not jeopardize extremely sensitive relationships with the rest of the world. This is precisely what they got. Considerations of security, safety and reliability were far more important than innovative applications.

Purkitt and Burgess argue that much of the pressure for weapons production came from South Africa's community of nuclear scientists rather than the military itself. However they also argue that nuclear weapons ultimately became attractive to senior members of the South African military and its closely related arms research and procurement community both for the weapons' enhancement of the country's military capability and for their contribution to increasingly sophisticated research and development of ancillary technology.[106] These arguments probably understate the political value of the weapons to South Africa's senior leaders and overstate their value to a military establishment never quite sure how the capability could effectively be employed. The South Africans seem to have intended their nuclear weapons to serve political purposes rather than military, and these purposes did not require the versatile inventories, multiple delivery options, fusion weapons or target lists of the great powers.[107] In any event, South Africa's resources were limited and significant additional investment in nuclear technology would almost certainly have compromised other equally important priorities.

Armscor, for its part, undertook the production of nuclear weapons as an engineering problem and was able to progress from a standing start to a deployable nuclear weapon within five years, a notable achievement. But perhaps more important was the ability to manage the disparate threads of weapons production and delivery systems despite the burden of international sanctions, great-power scrutiny and intense foreign pressure to prevent nuclear proliferation. At least in terms of the related delivery systems, there was considerably spin-off of this program that continues to benefit the South African arms industry. And, for a brief period, an African country had joined the "nuclear club."

The G-5/G-6 Gun Systems

South Africa's investment in modern, long-range artillery also received an intense boost by the country's intervention in Angola in 1975, following the sobering encounter with Soviet-produced weapons in the hands of Cuban expeditionary forces.[108] In contrast to South Africa's obsolescent World War II era guns, Soviet clients in Angola had three formidable artillery weapons that significantly outranged their South African opponents. The Soviet M-46 130mm gun could attack targets at ranges up to about 27 kilometers, while the excellent, mobile, rapid-firing D-30 122mm gun could quickly engage targets out to about 15 kilometers. Truck-mounted BM-21 multiple rocket launchers could saturate an area with 122mm rockets up to ranges of about 20 kilometers.

The NATO alliance had sought to counter these 1960s era Soviet weapons in the European Theatre with their own increasingly capable systems. For their armored forces, the NATO armies in the 1960s had begun to deploy 155mm howitzers as their mobile "medium" artillery. These self-propelled howitzers, with a standard firing range of just over 18 kilometers, typically were mounted in heavy, tracked armored vehicles that required significant, continuous logistical support. Yet even these NATO guns were vulnerable. They depended for their survival on armor protection, mobility, air cover and support by longer-range heavy artillery.

Prior to 1975 the South Africans had little incentive to invest in such technology, a situation that changed after their intervention in Angola. However, at this point they needed artillery suited both to the peculiar environment of the "bush war" and the features of their operational style. Even assuming it were possible, the South Africans could not remedy their artillery deficiencies simply by copying or buying the equipment then standard in NATO. Instead, they required artillery that would accurately outrange their opponents and be deployable over large distances and difficult terrain. Their guns had to be relatively inexpensive, logistically economical, accurate, mobile, robust and simple to maintain in austere environments. And they had to be survivable even if employed autonomously and in small numbers.

Any country has difficulty designing and producing major combat systems from scratch on short notice, and the South Africans did not attempt it with artillery. Instead, they took two shortcuts. First, they copied and product-improved the Soviet multiple rocket launchers, building a similar weapon with even greater range and lethality (the *Valkiri*). Then, they also sought out the best gun technology available in the West. The South Africans recognized the exploitable potential in the artillery research of Gerald V. Bull, a Canadian-born arms designer whose Space Research Corporation (based in Quebec) had for some years been experimenting with long-range gun designs and had pioneered significant innovations. Bull had been trying to interest the United States in his new designs, but despite his ground-breaking work was having difficulty achieving much tangible response. (One country that had taken advantage of Bull's earlier research was Israel.)

By 1975, Bull's most advanced design was his GC-45, 155mm gun.[109] Among its many novelties was a significantly greater barrel length than the NATO standard, new propellants that allowed heavier charges without endangering the weapon, and a newly designed extended range full bore (ERFB) projectile. At the same time, the Swedish arms industry had been testing novel "base-bleed" artillery ammunition that provided a significant boost in range. A system incorporating Bull's new gun that with all of these innovations had a stunningly greater potential range (up to 40 kilometers versus the 18 kilometers range of the standard 155mm NATO artillery at the time).

South African acquisition of the GC-45 technology was a story with many of the lurid elements of a modern adventure movie, a rapid and highly convoluted process.[110] Landgren says that the South Africans first approached the Israelis, who had for some years been using earlier versions of Bull's gun designs, then took advantage of (US) CIA connections to approach the Space Research Corporation directly, presumably in 1975 or 1976.[111] Armscor then purchased shares in the firm and sent Armscor personnel to the Canada to study the new technology.[112] McWilliams says that weapons prototypes subsequently were tested covertly on Antigua and shipped to South Africa.[113] By 1977, the South Africans had acquired all the technology necessary for their own artillery development program, and launched an intensive (and successful) effort to build a weapon suited to their unique requirements. Their version of the gun was first displayed in a 1980 parade, and by 1982 it was being offered for sale to foreign consumers. Armscor's chairman asserted that the gun was entirely produced in South Africa.[114]

South Africa's arms industry was challenged by more than just the research and testing of the new design. Production of the new weapon also required the rapid development of an industrial capacity to manufacture its technologically sophisticated components, including the gun tube, carriage, and novel ammunition. Armscor's Lyttleton Engineering (LIW) apparently did most of the gun system developmental work and produced the finished system, while the Armscor subsidiaries Naschem and Somchem (which subsequently produced ammunition) conducted much of the ammunition experimentation. There was little about the resulting artillery weapon that was truly new on the world scene, and the Austrian firm Noricum was soon producing a similar gun system that it exported widely, but South Africa's version achieved a brilliant blend of the most promising artillery technology in a system better suited to the country's requirements than anything it could have acquired elsewhere.[115] Arguably, the new gun also was the best system of its type in the world at the time of its introduction.

What South Africa got was a 14-ton, 155mm gun, light enough to be packed into a C-130 (Lockheed L-100/Hercules) transport aircraft for convenient air movement. A 10-ton truck already in the South African inventory (the Samil-100 6x6) towed the weapon on the ground, carrying its basic load of ammunition and 8-man crew. The gun system had a small diesel auxiliary power engine powering each of the gun's four, large wheels. This small engine also could propel the gun for short ground distances at speeds up to 15 kilometers per hour, and the entire gun system could be rapidly emplaced and displaced (an important survival feature on a modern battlefield). The wheeled carriage was optimized to accommodate the soft and sandy soils of southern Africa. With its associated truck transport, the gun could be towed at speeds up to 90 kilometers per hour on a road, and up to about 15 across country in rough terrain. It had a rapid rate of fire for a towed weapon of its size: up to three rounds a minute for fifteen minutes and up to two rounds a minute for sixty minutes. (This meant that a small, two-gun section or a four-gun battery could bring devastatingly effective fire on a target, and could then move quickly to a new site before it was threatened by enemy counter-battery fire.) And best of all for the South Africans, their new artillery could propel either a conventional high-explosive 155mm projectile with impressive accuracy to a range of 30 kilometers, or a base-bleed projectile with somewhat less precision to 39 kilometers. Unlike the equivalent NATO ammunition, all the different South African warheads were built to use the same firing tables, facilitating rapid target engagement. A single gun could attack targets throughout an area of almost 1,000 square kilometers. Neither the Soviets nor their Cuban and African clients had anything like it.

The towed G-5 gun was an impressive achievement, all the more so for the rapidity of its development and deployment. But the South Africans wanted a self-propelled version for greater combat mobility also suited to their operational style and environment. By the early 1980s they had constructed prototypes of a unique, 6x6, wheeled, self-propelled gun, and the new weapon was designated the G-6. It was first displayed in 1982 and formally entered South African service in 1988. The new gun weighed about 47 tons combat loaded, featured a crew of three to five men, and in its basic load carried forty-five projectiles and fifty propellant charges. Its powerful diesel engine gave it a road speed of up to 85 kilometers and hour and a cross-country speed of up to 30. It carried sufficient fuel for an astonishing unrefueled combat range of 700 kilometers. The robust vehicle had a good capacity to push through the vegetation native to southern Africa, and contained the typical South African modular features that facilitated quick swapping of parts for rapid maintenance.[116]

The G-5 and G-6 guns received their baptisms of fire in Angola in the 1980s. The G-5s apparently were first used on a large scale in repulsing an Angolan offensive against South Africa's ally Jonas Savimbi in September 1985.[117] However, they and the G-6s got their most substantial field test in a September 1987 Angolan offensive, against which the South Africans committed modest numbers of their ground forces. The new artillery played a crucial role, proving enormously effective in disrupting large-scale attacks by Soviet-organized Angolan and Cuban combined-arms units.[118] The remote southern African location of these events – and the lack of media coverage in the West – probably prevented the South Africans from receiving the credit they deserved for their achievement. But their Angolan and Cuban opponents had no illusions about the effectiveness of the new weapons. During the 1987 offensive they made increasingly desperate efforts to find and destroy the South African guns.

The country's interest in state-of-the-art artillery survived the political trauma of the 1990s, and these weapons now represent the most visible face of the country's military technology on the world market. In 2003, the South Africans unveiled the latest iterations of their advanced 155mm guns (the G5-52 and G6-52 models). These longer-barreled versions ("52 caliber" versus their older "45 caliber" standard) are paired with South African-designed, velocity-enhanced, rocket-assisted projectiles (V-LAP) extending the maximum range of the guns to 67 kilometers, an extraordinary accomplishment that still is essentially unmatched by competing designs. Automated loading provides very high rates of fire. The new South African artillery is fully computerized, including a state-of-the-art artillery target engagement system. It is designed for use in conjunction with a remotely piloted aerial reconnaissance vehicle.[119]

The G-6 offers a number of interesting possibilities, including effective roles for autonomous operations by very small numbers of guns. With the G6-52, one gun can provide artillery fire throughout an area of 1,700 square kilometers from a single firing point, can engage a target with up to six rounds impacting simultaneously, and can sustain a rate of fire of eight rounds a minute for all forty of its on-board projectiles. The gun system can be fired within 45 seconds of halting from the march and can move away from its firing position within 30 seconds of discharging its last round. The G-6 vehicle itself has significant combat range, excellent cross-country mobility, and state-of-the-art, automated navigation.[120] These features make it very competitive on a modern battlefield.

While the sophistication of the weapons is worthy of note, equally remarkable are the crosscutting relationships within South Africa's arms industry that produce this system. Various Armscor entities were involved in translating the original military requirement into a functioning system, including product design and prototype testing. The basic gun system, turret assemblies and ammunition – projectiles and propellants – now are produced by several Denel subsidiaries in its Land Systems Group. Denel's Aerospace Systems offers the compatible *Seeker* II remotely piloted aerial vehicle.[121] The private sector enterprise African Defence Systems (ADS) provides the state-of-the-art AS-2000 artillery computer while another private sector enterprise, BAE Land Systems OMC, builds the G-6 chassis and vehicular components. The efficient coordination of a wide range of different industries to produce a well-integrated, high-quality product of this type is a South African specialty and a marketable asset.

Rooivalk

South Africa's military experiences in the 1970s and 1980s strongly suggested the usefulness of a combat helicopter that could be used in roles such as reconnaissance, aerial artillery support of ground forces and convoy escort. Like the US Army in the early 1960s and the Rhodesians in the 1970s, the South Africans made some effort to fill this need by arming their utility helicopters.[122] But by the 1970s, they also had observed the development of much more capable attack helicopters like the US Army's AH-1 *Cobra* and the Soviet Mi-24/*Hind*. They also had encountered the Soviet-designed *Hind* in action in Angola.

The South Africans had long contemplated the military requirement for such a weapon when they unveiled their first attack helicopter prototype, the *Alpha* XH-1, in March 1986. This machine, built by Armscor's Atlas Aircraft Corporation, appeared to be highly modified *Alouette* III.[123] However, even as they displayed the experimental XH-1, the South Africans seem to have been contemplating a larger, more capable aircraft that exceeded the capabilities of the Soviet-designed *Hind*, so the XH-1 apparently served mainly as a test bed. Atlas had commenced serious developmental of a new combat helicopter by the end of the 1980s, selecting the French-designed *Puma* as its basis. (The *Puma* was by then South Africa's main utility helicopter and locally manufactured.) The South Africans largely had perfected their new design by the mid 1990s. Designated the *Rooivalk* (Red Falcon), it was first publicly displayed at the UK Farnborough Review in 1996, and entered service with the South African Air Force in 1999.

The combat helicopter that emerged from these efforts was a highly capable machine, smaller and lighter than the Soviet-designed *Hind* but somewhat larger and heavier than the US Army's AH-64D *Apache Longbow*.[124] Like the *Apache*, it was designed for a two-man crew of pilot and weapons officer (and in this respect it was unlike the *Hind*, which was designed for transport of up to eight troop passengers). While the *Rooivalk* lacked the *Hind's* speed, range and payload, it had significantly better handling qualities and protective features. Its cruising speed and rate of vertical climb equated to the *Apache Longbow*, while its combat range was about three times that of the latter (both equipped with external fuel tanks).

The *Rooivalk's* lineage was very apparent – the reliable *Puma* drive train and uncommonly robust rotor and rotor-head assembly contributed to impressive qualities in adverse environmental conditions. Two state-of-the-art Turbomeca Makila 1K2 turboshaft engines powered the aircraft.[125] It had a maximum speed of 167 knots, a fast cruise speed of 150 knots, a maximum sideways speed of about 45 knots, and a maximum rate of climb of 2,200 feet per minute.

Its combat range was an impressive 1,130 kilometers (with external fuel cells). Military pilots from Western countries who have flown it generally have given it very high marks for its handling qualities. The aircraft also was designed for battlefield survivability, reflecting efforts to minimize infrared signature, visibility and noise. A very sophisticated self-protection system includes radar warning, laser warning and countermeasure systems of flares and chaff. Its designers also gave considerable thought to crash-resistant features.

The *Rooivalk's* battlefield function was supported by up-to-date navigation systems and an automated flight control system. These aids included a Doppler radar velocity sensor, global positioning system, heading sensor and air data unit, all of which provided data both to the crew and to the flight control system. That system, in turn, conducted automatic rate stabilization, attitude and altitude hold, turn coordination, hover capture/hold and trimming. Vehicular performance was matched by sophisticated target detection, acquisition and tracking capabilities and by a versatile mix of state-of-the-art weapons produced in South Africa. A nose-mounted, stabilized sight contained a television monitor, forward-looking infrared sensor, laser rangefinder and laser designator. The two-man crew was equipped with helmet-mounted sight displays slaved to the aircraft's weapons providing a "head-up" information display for nap-of-the-earth flight.

The *Rooivalk's* weaponry was impressive, though roughly equivalent to other top-of-the-line attack helicopters. It had a chin-mounted, rapid-firing 20mm cannon. Other armament could be suited to the customer and the mission, including a mix of anti-tank missiles, anti-aircraft missiles or 70mm folding-fin aerial rockets. The *Rooivalk* could employ the US *Hellfire* or European HOT 3 anti-tank missiles, but its standard anti-tank armament was the Denel (Kentron) *Mokopa*, with a choice of semi-active laser or MilliMetre Wave guidance. It could carry up to sixteen anti-tank missiles and up to four air-to-air missiles, typically the MBDA *Mistral*.

In 2005 the *Rooivalk* still was the flagship product of Denel Aerospace Systems (the contemporary successor of Atlas Aircraft) and remained competitive in capability and pricing to the most advanced attack helicopters produced elsewhere. South Africa had engaged in vigorous efforts to sell it abroad, though without much tangible success. An early, promising prospect of a Turkish sale was scuttled by South Africa's concerns over Turkey's human rights record. The Malaysian military had expressed interest, but (as this was written in 2005) apparently lacked the financial resources to invest in the helicopter. The South African Air Force operated the only *Rooivalks* in active service – a total of twelve.[126]

Given the limited foreign interest, Denel had not been able to recoup the development investment in this aircraft, and by 2004 there were calls in South Africa simply to abandon the *Rooivalk* as an expensive albatross. Meanwhile, industry insiders displayed frustration that a machine with its obvious merits did not compete more effectively in the international arms market. But nations that buy foreign weapons often are motivated more by the political relationships that come with an arms purchase than by the details of the equipment itself, a factor that favors the industries of the more prominent industrialized nations. Then, given the high costs of military aircraft for any country, domination of the helicopter markets by established Western industries, and prevalent suspicions that the South Africans may not be able to service and upgrade their materiel as effectively as the Western firms, it probably is not surprising that export sales have been lacking. South African industries may even have inadvertently undermined the prospects for *Rooivalk* sales by their ability to upgrade older Cold War systems with modern components. Still, the creation and fielding of the *Rooivalk* is a remarkable achievement matched by few other countries.

The SMR 95 Jet Aircraft Engine

Even before the end of UN sanctions in 1994, South Africa's commitment to political reform had given it previously inconceivable access to the international arms market. An early South African effort to take advantage of the new political environment was a project to upgrade the capabilities of its combat aircraft (particularly the Mirage F-1 fighter) with a sophisticated new engine. The South Africans chose an engine originally designed in the Russian Klimov Plant for the top-of-the-line Soviet MiG-29 interceptor. (Klimov called the engine the RD-33 and the South Africans designated their modification the SMR 95.) This project was intended to rectify a key South African military vulnerability – the comparatively anemic power plants of South Africa's aging jet fighters.[127] The private sector South African firm Aerosud performed the technical modifications of both engine and the aircraft under the direct supervision of the South African Air Force.

The project continued from 1990 to 1994, and included tests of the new engine in the Mirage F-1 and *Cheetah*. (These tests simultaneously evaluated a wing modification on the *Cheetah*.) By 1994, the South Africans had completed the flight tests for the Mirage F-1 and were beginning the tests on the *Cheetah*, all indicating a highly successful conversion and significantly enhanced aircraft performance. The conversion provided radical improvements in the F-1's payload and turning performance, and a quantum increase in the *Cheetah's* payload. However, the political changes in South Africa in 1994 resulted in massive redirection of government funding and the project was terminated.[128] That decision was unfortunate. A continuation of the project could significantly have extended the useful life of the F-1s, which by that point largely had been withdrawn from active service. It could also have graphically demonstrated South Africa's technical expertise in upgrading and modernizing older military equipment. Ironically, the *Defence Review* of 1998 reestablished a South African requirement for a state-of-the-art fighter aircraft, but the ultimate choice of aircraft was to be strongly influenced by the counter-trade offers associated with the Defence Packages. The new fighter would be the Swedish-designed *Gripen,* not an upgraded F-1 or *Cheetah*.

While the South African government did not elect to pursue the engine upgrade of its older jet fighters, the successful modification and the ability to marry a third generation Russian fighter engine to two different second generation French-designed aircraft was a notable achievement in aviation engineering. It reflected an entrepreneurial problem-solving approach more reminiscent of the American frontier than the bureaucratic, stolid aerospace industries of the contemporary developed world. (By way of analogy, in 2005 it was hard to imagine a small team of US Air Force supervisors and Boeing engineers working to match Chinese engine technology to British airframes to get radically enhanced aircraft performance for the US Air Force.)

Some Implications of the "Benchmark" Products

Beyond the nuclear weapons program, the long-range artillery, the *Rooivalk* helicopter and the SMR-95 engine, a variety of other military products are viable "benchmark" indications of the South African capability to produce sophisticated military materiel. These range, across a spectrum of ground, aerospace and electro-optical applications. They could include, for instance, the frequency-hopping tactical radios developed in the 1960s, the pilot helmet sight slaved to air-to-air weaponry that appeared in the 1970s, the signals intelligence/electronic warfare equipment deployed in the 1970s and 1980s, the impressive *Rooikat* reconnaissance vehicle that appeared in the 1980s, the incipient space program of the 1980s or the air-to-ground "smart" munitions of the 1980s and 1990s. And even this list fails to mention a variety of other South African accomplishments in the conception, design and deployment of advanced military materiel from the early 1960s to the present.

Whatever list may be offered, the benchmark products point to at least two key features of the South African arms industry. One is its sophistication, adaptability and versatility. For a "mid-sized" country in the developing world, South Africa has produced an astonishing variety of military materiel. Other developing countries have sophisticated arms industries – India and Brazil come immediately to mind – and a handful of other developing countries have military space programs and nuclear weapons. But with the conspicuous exception of Israel or China, it is difficult to match South Africa's versatility except in the arms industries of the developed West. And the South Africans have regularly produced military equipment with capabilities superior to analogous materiel in the inventories of the developed world. Perhaps every bit as significant is South Africa's demonstrated ability to take a pressing military need and rapidly find a solution compatible with the country's technological capabilities, the operational environment and the South African military culture.

These features could be dangerously destabilizing in the hands of a rogue regime or an aggressive leadership bent on coercing the country's neighbors. Yet there are no indications that South Africa will pursue such courses in the foreseeable future, and the traumas of the 20th Century seem to have provided a salutary immunization against them. It is more useful to view the South African arms industry simply as a surprisingly versatile tool among the various instruments of power available to one liberal democracy in the developing world.

South African Military Technology

A key question still remains: just how effective is the security-related technology and materiel available from South African sources? This question is easier to pose than to answer, but it is useful to distinguish several related issues. First is the difference between technological capacity and produced materiel: in 2005 South Africa had more of the former than the latter. Another is the distinction between truly novel technology and clever adaptation of existing technology: again, South Africa has shown much more proclivity for adaptation than for true novelty. Still another issue is the capacity to partner in ways that allow rapid combinations of new and older technology in the same systems. The South African advantages are more in capability and potential than fielded materiel and more in adaptation and partnership than truly "cutting edge" technology. The South Africans themselves have identified "core products" they consider competitive on the world market,[129] and others they deem economically viable and worth a continuing investment.[130] But their listing alone does not do justice to their capabilities.

A significant characteristic of South African industries is a willingness to tailor products to the needs of specific consumers, even for limited product runs. In the 1970s and 1980s, the defense industries typically produced small quantities of specialized items for the South African military, a feature not generally characteristic of the larger defense industries in the industrialized countries, with the latters' emphasis on economies of scale. The lack of economy of scale in the arms industry may have been one of the roots of the national economic malaise in the waning years of the *apartheid* era, but customer responsiveness was a feature much appreciated by military clients. South African defense industry spokesmen still express an eagerness to tailor products to the needs of consumers.[131] That tailoring cuts across industries (and across public and private sectors).[132]

The South African capacity to field military materiel has been significantly facilitated by the development of "world-class" ranges and test facilities on a scale astonishing for such a small country in the developing world. Subsidiaries of Armscor and Denel still manage the most sophisticated of these facilities.[133] (See Figure 4.5 and Appendix A for additional detail.) Private sector defense industries and the military services also maintain facilities for testing their own equipment.[134] Armscor and Denel use the facilities for their research, provide access to other South African companies, and have had some success marketing them to European and Asian defense industries. Any assessment of South African defense-related technology would be obliged to acknowledge the advantages offered by the diversity and sophistication of the test facilities. These can be rented by externally based industries, but they also provide to a continuing capacity for rapid development and testing of prototypes for South Africa's own use.[135]

In addition to the excellent research and development facilities, South Africa has a surprisingly large number of organizations (and individuals) that might be described as "technology enablers." These include sophisticated consultants on security issues and enterprises seeking to serve as interfaces between military consumers and developers or producers. "Enablers" particularly include experts in cyber applications. South Africa's socio-economic elite embraced cyber technology very early, and adapted it on a large scale to security needs during the latter years of the *apartheid* era. Today, a number of South African companies provide sophisticated cyber equipment, software, simulations, command and control, other applications or advice to military and civilian consumers.

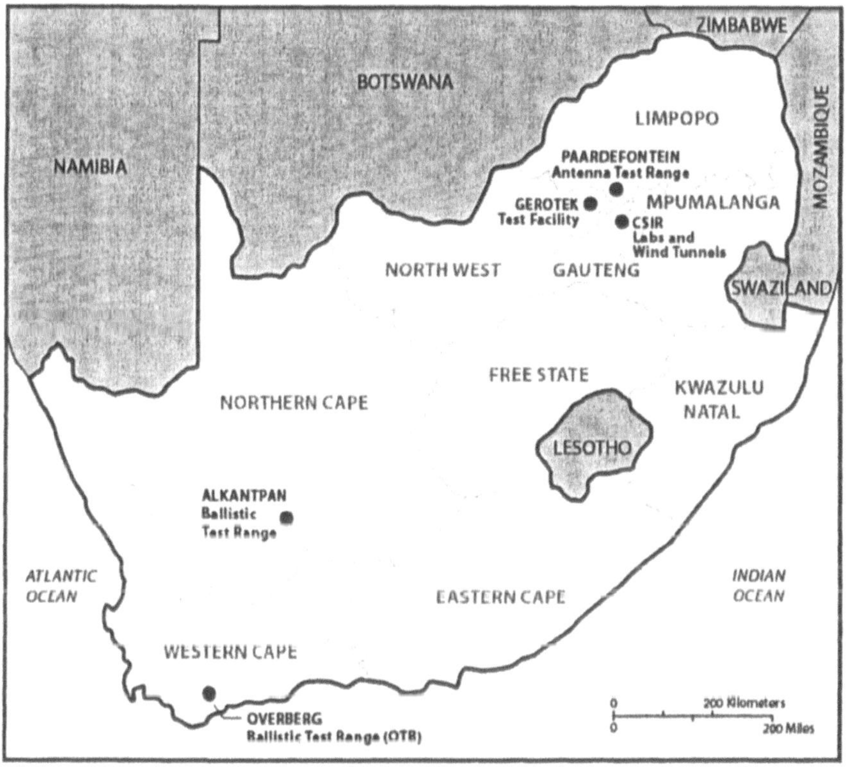

Figure 4.5. South African Arms Industry Testing Facilities

The South Africans still place considerable emphasis on modularization of military systems. This partly is a heritage of the *apartheid* years under UN sanctions, when the defense industry could not produce everything it needed and was obliged to buy certain items abroad that could quickly be adapted to an existing need. This also partly is a response to an operational environment of long distances, primitive infrastructure and austere logistics capabilities: South African battlefield mobility once depended on the ability to rapidly "swap out" damaged components.

The emphasis on modularization is enhanced by an arms industry very comfortable with extensive subcontracting, which on relatively short notice can produce tailored "packages" of components such as gun systems or communications and electronic warfare systems for installation in consumers' major end items. The industry can apply the same modularization to components subcontracted from foreign sources, or to weapons systems purchased from foreign suppliers (as evident in the subcontracting for the Strategic Packages).

When it comes to defense equipment itself, the South Africans typically have designed their military materiel for the rigors of the southern African environment. That environment poses a variety of challenges, including long distances and primitive infrastructure, varying altitude, rough terrain, extremes of heat, cold, aridity, humidity and dust, and heavily mined areas. Efforts to meet these challenges resulted in materiel that is relatively simple to maintain, robust, able to negotiate restrictive terrain rapidly, economical, functional in environmental extremes, and protected from low-technology threats. These remain very useful characteristics for a variety of military applications. While the technology generally has not broken new ground, it reflects a clever adaptation of proven concepts and materials.

There are a few limited areas of fielded military technology in which South African materiel is "cutting edge" and many in which it is state-of-the-art. One area for the former is that of tactical communications/electronics, electronic warfare and electro-optical applications, a legacy of strenuous South African efforts in the *apartheid* era to maintain an advantage over Soviet-equipped opponents. This particularly was true in combat aircraft: the South Africans could not match the advanced aerodynamic and engine technology of later Soviet aircraft designs, but they endeavored to compensate with advanced avionics, missiles and electronic warfare. South African companies achieved sufficient technical sophistication in these fields that after the advent of majority rule, they could maintain some advantage and quickly establish partnerships with European defense industries to sell their technology on the world market in European-produced aircraft and vehicles. The European partnerships now assure that South Africans can keep up with latest advances in communications/electronics and electro-optical applications, and can compete effectively in niche applications such as miniaturization. South African industry remains competitive across a spectrum of tactical communications and electro-optical equipment.

Another South African "niche" advantage is in the design and production of wheeled combat and security vehicles. South Africa's conflicts in the 1970s and 1980s encouraged the local development of vehicles that combined long range, robustness, modularization for rapid field repairs, speed and mobility in difficult terrain, protection from mines and capacity to be reconfigured for specialized roles. The South Africans developed a family of transport trucks, specialized mine-protected vehicles, multi-role wheeled combat vehicles, an armored reconnaissance vehicle, and the chassis for a self-propelled gun. At the time of their introduction from the mid 1970s to mid 1980s, each of these systems was state-of-the-art. Although the original versions now are dated, South African industry retains the capacity to produce upgrades and follow-on systems that preserve many of the advantages of the originals. Here, however, the South Africans probably will depend on foreign partnerships for some technology: it is not clear, for instance, that domestic industry will ever find it cost effective to produce the necessary vehicular engines, though it is capable of doing so. South African companies also have developed specialized vehicles for a variety of security-related roles and have had some success in marketing them. BAE Land Systems OMC, for instance, has sold multi-role environmentally protected vehicles to several law enforcement organizations and to the military in North America.[136]

Related to vehicle protection is the fact that South African industries are widely recognized as world leaders in production of mine-resistant and environmentally protected technologies, a key feature of their latest ground combat vehicles. During the *apartheid* years, the South Africans required greater counter-mine capability than was available elsewhere. As a result, they invested in research, development and fielding of technology across a spectrum of design, materials and modularization. They are likely to remain competitive in these technologies. South Africa is increasingly visible for its de-mining expertise and has fielded some very effective equipment now used around the world in this role.[137] The Denel subsidiary Mechem has devised a holistic mine detecting process, including the use of dogs, that genuinely is state-of-the-art. South Africa's expertise in demining and manufacture of mine-protected vehicles is evident in the enthusiasm of the United Nations for South African equipment.[138]

Another area in which the South Africans themselves recognize a special capability is in the production of tactical weapons systems, to include missiles, artillery, projectiles and explosive ordnance. Their missile systems are, in general, state-of-the-art and their growing partnerships with US and European armaments industries probably will assure that they maintain interest and expertise. South African research and development has pushed into some novel areas: for instance, in the development of ultraviolet guidance for air-to-air missiles to attenuate the effect of background "noise." The South African military still is the most significant consumer of its industry's missiles (and, not surprisingly, these feature prominently in the domestic production for the Defence Packages). However, in the long term, it is unclear whether the local military will remain a large-scale consumer. The viability of this South African technology depends on exports to peer militaries outside of Africa, and export options are profoundly influenced by the external partnerships.

Artillery has been something of the "poster child" for South African arms exports since the 1980s, when the country began selling its G-5 155mm extended range gun system to clients like Iraq. The G-5 represented a quantum leap in ground-based indirect fire capabilities for South Africa, and it also reflected state-of-the-art technology. Its advantages were not so much the brilliant technological advances of the individual components – gun tube and chassis, fire control, base-bleed extended range ammunition or warhead lethality, but the synergy of having the best of each in one highly compatible system. The entire package was so good that twenty years after first fielding, with product improvement, it still is as advanced as any equivalent system on the market. The South Africans seem determined to remain competitive in field artillery – their innovative new G7 light 105mm gun/howitzer is likely to be an industry leader in this niche. However, more interesting than the guns themselves is the industry that can design and produce them.[139] South African firms have no difficulty conducting the research, testing and assembly necessary to field the most sophisticated gun systems.

While communications/electronics, electronic warfare, instrumentation equipment, mine-resistant combat vehicles, and tactical crew-served weapons are areas of particular expertise, the South Africans are highly capable of servicing smaller niche markets. They largely backed out of the military small arms production after 1999 due to market saturation, but they continue to produce small arms and specialized military and police weapons such as special purpose sniper rifles like the Denel (Mechem) 20mm gun that stirred US military interest in 2004 and 2005. Several South African firms are conducting research on non-lethal weapons and it seems likely that South Africa could compete in that market. South African firms also have researched areas such as specialized combat uniforms that attenuate the infrared signature of the human body. And the South African have both the research and manufacturing capability to market hazardous materials protective systems in quantity.

Botha calls attention to a particular South African expertise in systems design, systems engineering, systems integration and equipment upgrading.[140] The South Africans are very good at taking state-of-the-art components from a variety of industries and integrating them into tailored packages in order to enhance the performance of Cold War era weapons. They have successfully marketed upgrades for older Soviet equipment like the ZSU-23-4 self-propelled air defense gun, the T-72 main battle tank, BMP infantry fighting vehicle and the Mi-24/*Hind* attack helicopter. They seem willing to perform these upgrades on a smaller scale than their potential competitors in Europe or North America. South African technology for such upgrades probably is at least equal to, and may well surpass, that available from other potential competitors.

It seems unlikely that South Africa will achieve a dominating role in any domain of armaments production in the foreseeable future – nor has it expressed any intention to do so. However, it is poised to play effective niche roles. It has a proven capability to compress the R&D and production cycles in areas of particular expertise and can offer quality products with little lead-time. Its defense industries tout their ability to tailor products to unique customer requirements to a greater degree than the competition. And South African researchers have a well-refined appreciation for the unique challenges of the southern African operational environment: for forty years, they have adapted and innovated military technology to suit that environment. It is unlikely that anyone else can best them in this capability.

At this time, South Africa's arms producers still look to their country's military as a prime consumer, but have made efforts (profound in some cases, less so in others) to find external customers. They have succeeded in some overseas markets, including some in North America that are notoriously hard to penetrate. They have demonstrated an ability to manufacture high-quality, high-technology products. Partnerships with European arms producers have been the most lucrative enabler of external sales to date, but the South Africans are seeking additional partnerships and connections. The second decade after majority rule probably will be a propitious time for external industries to establish relationships with their South African counterparts.

Notes

1. Helmoed-Römer Heitman, personal communication with the author, 12 June 2004; and interview by the author, 23 June 2004.

2. The foundational legislation was the Armaments Development and Production Act, 1968, (Act No. 57 of 1968) subsequently amended by other acts of Parliament.

3. Armaments Corporation of South Africa, Limited Act, 2003, Section 2, http://www.info.gov.za/documents/acts/index.htm#2003, (21 March 2005).

4. This board included the CEO of Armscor, the Secretary for Defence (senior civil servant in the Department of Defence) and the Chief of the South African National Defence Force. Other board members were drawn from the private sector, though one was a former Chief of the South African Defence Force. The eleven-person board included three women (two of whom resigned in 2003) and two white South Africans. ARMSCOR *Annual Report*, 2003-2004, 12-13, http://www.armscor.co.za, (21 March 2005).

5. Armscor also served as the acquisition agency for the South African Police Service and Department of Correctional Services. In the future, it may play a similar role for external clients such as the African Union, Southern African Development Community or even other African states. ARMSCOR *Annual Report*, 2003-2004, 20.

6. ARMSCOR *Annual Report*, 2003-2004, 1.

7. Armaments Corporation of South Africa, Limited Act, 2003, Section 4. See also *White Paper on Defence Related Industries*, Chapter 3, paragraphs 35, 36 http://www.info.gov.za/documents/whitepapers/index.htm#1999, (2 March 2005).

8. ARMSCOR *Annual Report*, 2003-2003, http://www.armscor.ac.za, (15 June 2004); ARMSCOR *Annual Report*, 2003-2004.

9. For a description of Denel's separation from Armscor, see Peter Batchelor and Susan Willett, *Disarmament and Defence: Industrial Adjustment in South Africa* (Oxford University Press, 1998), 84-101. The new 2003 legislation now has specifically empowered Armscor to seek additional commercial opportunities so it could conceivably engage in future production and more aggressively market its research and testing resources to private sector actors. See Armaments Corporation of South Africa, Limited Act, 2003, Section 4.

10. For additional data, see http://www.armscor.co.za/DefenceScienceTechnology/ListCompanies.asp, (25 March 2004).

11. The show in September 2004 was staged at Waterkloof Air Base in the Tshwane (Pretoria) area. See http://www.aadexpo.co.za, (15 March 2005) and ARMSCOR *Annual Report*, 2002-2003, 5.

12. Armaments Development and Production Act, 1968, (Act No. 57 of 1968) subsequently amended by other acts of Parliament.

13. The new legislation was the Armaments Corporation of South Africa, Limited Act, 2003 (Act 51 of 2003), signed into law by the State President on 20 April 2004; see also ARMSCOR *Annual Report, 2003-2004*.

14. ARMSCOR *Annual Report, 2003-2004*, 20. The treaty documentation that establishes the Southern African Development Community (SADC) Organ on Politics, Defence and Security makes provisions for a Mutual Defence Pact that seemingly would facilitate use of Armscor's acquisition expertise by other African signatories. Armscor also has taken formal note of the African Standby Force discussed in the Peace and Security Council of the African Union, believing that Armscor could play an acquisition management role for that entity.

15. Armscor derives limited income from commercial activity, but was prohibited by the terms of the 1992 separation (from Denel) to engage in non-defense production *per se*, severely limiting its access to other sources of income. However, the new legislation signed in 2004 has specifically empowered Armscor to seek additional commercial opportunities so it could conceivably engage in future production and more aggressively market its research and testing resources to private sector actors, including those outside the defense sector. See Armaments Corporation of South Africa, Limited Act, 2003, Section 4.

16. Armscor's workforce is ageing – a concern in itself – but another substantial problem is the difficulty of recruiting the diverse workforce reflective of the new South Africa. Still, the organization apparently is making strenuous (and at least modestly successful) efforts in this area. ARMSCOR Annual Report, 2003-2004, 3, 8.

17. ARMSCOR *Annual Report, 2002-2003*, 5, 10-1.

18. The actual transfer payment increased by 19% between 2002-2003 and 2003-2004 (from R209.1 million to R249.64 million). Previous years' annual increases had averaged 5.3 percent. ARMSCOR *Annual Report, 2003-2004*, 15.

19. The new Act does allow Armscor to "exploit commercial opportunities" (Armaments Corporation of South Africa, Limited Act, 2003, section 4) but it still is unclear if future Defence allocations will be sufficient to make up the difference between commercial income and current shortfalls.

20. Information provided by Alan Holloway, Vice Chairman of AMD and Managing Director of ANSYS, private communication of 4 April 2005. The author speculates that the new white paper will endeavor to define Armscor's new prerogatives authorized by the Armscor Act of 2003.

21. *Denel Annual Report 2002* http://www.denel.co.za, (1 July 2003). See also Batchelor and Willett, *Disarmament and Defence*, 84-101.

22. *Denel Annual Report*, *2004, 2* http://www.denel.co.za, (1 March 2005).

23. The Chairman of the board was an investment banker. Board members were all civilians, but included two retired generals. Two members were women. All but four were nonwhite. *Denel Annual Report*, 2004, 8-9.

24. Ibid., 18-9. Executive Committee members were functional experts, typically with long experience in the industry. They included two women and six nonwhites.

25. Ibid., 20-31.
26. Ibid.
27. Ibid.
28. For details, see David Botha, *South Africa's Defence Industry: Charting a new Course?* (Pretoria: Institute for Security Studies, ISS Occasional Paper 78, 2003), 9-10. For a more detailed (but dated) historical treatment, see Peter Batchelor and Susan Willett, *Disarmament and Defence: Industrial Adjustment in South Africa* (Oxford: Oxford University Press, 1998), 84-101.
29. The 2004 *Annual Report* indicated a total of 10,925 employees. *Denel Annual Report*, 2004, 2; but see also Khulu Phasiwe, "Denel Primes Itself to Offload Workers," (Johannesburg) *Business Day*, 10 February 2005.
30. Ibid.
31. See, for instance, Helmoed-Römer Heitman, "Opportunity Knocks," *Jane's Defence Weekly*, 3 August 2005, 28.
32. *Denel Annual Report*, 2003, 24; *Denel Annual Report*, 2004, 6. In conferences with US officials, Denel's representatives have indicated that from 1994 to 2004, their production emphasis has shifted from 90 percent domestic sales and 10 percent exports to the exact reverse, an amazing transformation. Information provided by Duncan Lang, contractor to the US Department of Defense, 11 March 2004.
33. *Denel Annual Report 2002*, 15, 36-68; *Denel Annual Report*, 2004, 52.
34. Information provided by Duncan Lang, contractor to the US Department of Defense,12 March 2004. See also *Denel Annual Report*, 2004, 52.
35. Peter Batchelor and Paul Dunne, "Industrial Participation, Investment and Growth: The Case of South Africa's Defence Related Industry." (Paper prepared for the Centre for Conflict Resolution, Cape Town and Middlesex University Business School, United Kingdom, February 2000), 4.
36. Denel's 2003 *Annual Report* attributed its 2003 loss primarily to the strong Rand, which adversely impacted profits. *Denel Annual Report*, 2003, 20. The 2004 *Annual Report* continued to complain about the strong Rand relative to the US dollar, but also mentioned Denel's "one off" acquisition in 2003-2004 of a significant debt obligation. *Denel Annual Report*, 2004, 51.
37. *Denel Annual Report*, 2003, 24; *Denel Annual Report*, 2004, 51.
38. Botha, *South Africa's Defence Industry*, 10-2. See also Heitman, "Opportunity Knocks," 24, 29.
39. Most of CSIR's focus has always been on non-military issues, with current interests ranging from the environment to mining to building technology to advanced cyber applications. For organizational details, see http://www.csir.co.za, (13 April 2004).
40. Ibid.
41. Author's interview with Andre Nepgen, Director, Defencetek, 20 June 2003.
42. See http://www.csir.za, (13 April 2004); Nepgen interview, 20 June 2003; information provided by Duncan Lang, contractor to the Office of the US Secretary of Defense, 12 March 2005.
43. Nepgen interview, 20 June 2003. Among the achievements listed in its literature are applications in continuous force testing technique in medium speed wind tunnel (MSWT); flight tests on new generation DAP flutter exciter for civilian aircraft; linux-based parallel clusters for computational fluid dynamics; Enigma II electronic countermeasure and radar target simulator; ground-based air defense systems; infra-red electronic warfare; and digital mortar sight linked to a hand-held ballistic computer.

44. Ibid.

45. Batchelor and Willett, *Disarmament and Defence*, 38.

46. See http://www.ads.co.za, (27 April 2005). Helmoed-Römer Heitman, *Unpublished Manuscript* provided to the author in June 2003, 141. A black-owned South African investment company, Futuristic Business Solutions, holds a 40% share.

47. Additional details are available at http://www.ads.co.za, (27 April 2005) and http://www.amd.org.za, (12 May 2004).

48. For details, see http://www.amd.org.za, (12 May 2004) and http://www.world security-index.com/details.php?id=265, (12 May 2004).

49. For details, see http://www.ate-aerospace.com, (15 May 2004) and http://www.sadid.co.za/SADID_7/edition7/ate/ate.html, (15 May 2004).

50. Ibid. Heitman notes that this enterprise is new to the difficult military ground vehicle market and he is somewhat skeptical of its prospects. Helmoed-Römer Heitman, private communication with the author, 12 June 2004.

51. The country in question was Algeria. Heitman, personal communication, 20 April 2005.

52. Gerber interview, 4 June 2003 and personal communication with the author, 4 June 2004 and 7 June 2004; Heitman, interview with the author, 23 June 2004. Ironically, the South African Air Force could have benefited from such an upgrade to its own Mirage F1s, reducing the immediate need for a new multi-role fighter, but elected instead to decommission the F1s in its inventory.

53. By 2005 ATE had performed this upgrade for Mi-24/*Hinds* of the Algerian Air Force. Heitman, personal communication. 20 April 2005. See also http://www.sadid.co.za/ SADID_7/edition7/ate/ate.html, (15 May 2004). Integrating systems that it contracts from Denel and other suppliers, ATE offers a "*Superhind*" upgrade of the Mi-24 that can include new South African missiles, a turreted 20mm automatic cannon, vastly improved target acquisition equipment, improved navigation systems and a night-fighting capability.

54. Information provided by Duncan Lang, contractor to the Office of the US Secretary of Defence, 11 March 2004.

55. Author's interviews with Reginald Ngwenya, Chairman of AMD, Alan Holloway, Vice Chairman of AMD and Managing Director of ANSYS and Paul J. Gerber, Group Manager for Business Coordination and Counter-trade, Grintek Group Limited, 21 June 2004.

56. David Botha, *Offsetting the Costs of SA's Strategic Defence Packages* (Pretoria: Institute for Security Studies, ISS Occasional Paper 75, 2003), 7.

57. Information provided by Paul J. Gerber, Group Manager for Business Coordination and Counter Trade, Grintek Group, Limited (who served as Director of the South African Air Force Fighter Programme at the time), personal communication with the author, 4 June 2004. Additional details on this program are provided later in the chapter.

58. See http://www.aerosud.co.za, (20 May 2004); Nic Dawes, "No plane, no gain," (Johannesburg) *Mail and Guardian*, 9 December 2004.

59. See http://www.baesystemsomc.co.za, (26 March 2005).

60. The South African investors are DGD Technologies (Pty) Ltd, a Black Empowerment Enterprise (BEE), http://www.baesystemsomc.co.za, (26 March 2005).

61. Author's interview with Brigadier Tony Savides (South African Army, retired), Director of International Marketing, Alvis-OMC (now BAE Land Systems OMC), 18 June 2003.

62. The acronym "OMC" recalls the firm's origin as the *Olifant* Manufacturing Company, an enterprise set up by the South Africans in the 1970s to upgrade their older

British-made *Centurion* main battle tanks into the *Olifant* model.

63. Savides interview, 18 June 2003; http://www.baelandsystemsomc.co.za, 26 March 2005.

64. See http://www.grintek.com/, (15 May 2004).

65. Heitman notes the prominence of Kunene Brothers investment in the South African-owned share, sufficient to make Grintek itself a Black Empowerment Enterprise (BEE). Heitman, personal communication, 12 June 2004.

66. Ian Scott, ITWeb (Johannesburg) News, 22 February 2005.

67. See http://www.reunert.co.za, (22 May 2004).

68. Ibid; Heitman, personal communication, 20 April 2005. The US Navy's Littoral Combat Ship reflects a relatively new concept in US naval thinking. In May 2004, the Navy contracted General Dynamics and Lockheed Martin each to produce two prototypes of the new ships, and may ultimately acquire up to 60 vessels of the successful design. Series production probably will commence in 2009. The Lockheed Martin prototype includes the EADS TRS-3D C Band Radar with the Reunert components built in South Africa. For more information on the Littoral Combat Ship, see http://www.naval_technology.com/projects/littoral, (28 April 2005).

69. See http://www.armscor.co.za, (26 March 2005) for a more comprehensive listing and description.

70. Botha, *South Africa's Defence Industry*, 12.

71. Ibid.

72. See http://www.ccii.co.za, (15 May 2004) for detail.

73. Heitman, *Unpublished Manuscript*, 150; http://www.aktelux.com/article_2005_01_11_1135.html, (28 April 2005).

74. See http://www.ams.co.za, (12 May 2004) for detail. Also based on information provided by Heitman, personal communication, 12 June 2004.

75. See http://www.safair.co.za, (20 July 2004).

76. Author's interview with Johan van Rensburg, Commercial Specialist, US Commercial Service, 23 June 2004; Heitman interview, 23 June 2004.

77. The author is indebted to South African industrialist Alan Holloway for suggesting this discussion and offering an authoritative recommendation on the items that should be included in it.

78. South Africa's nuclear weapons program has produced a rich and interesting literature, though a number of questions about it remain unanswered. The most authoritative recent treatment is Helen E. Purkitt and Stephen F. Burgess, *South Africa's Weapons of Mass Destruction* (Bloomington, IN: Indiana University Press, 2005). For other useful treatments, see Hannes Steyn, Richard van der Walt and Jan van Loggerenburg, *Armament and Disarmament: South Africa's Nuclear Weapons Experience* (Pretoria: Network Publishers, 2003); Roy E. Horton, III, *Out of (South) Africa: Pretoria's Nuclear Weapons Experience* (Colorado Springs, CO: USAF Institute for National Security Studies, Occasional Paper 27, 1999); Peter Liberman, "The Rise and Fall of the South African Bomb," *International Security* 26, no.2 (Fall 2001): 45-86; Waldo Stumpf, "South Africa's Nuclear Weapons Program: From Deterrence to Dismantlement," *Arms Control Today* 25, no. 10 (December 1995/January 1996); and D. Howlett and J. Simpson, "Nuclearisation and denuclearization in South Africa," *Survival* 35, no.3 (1993): 154-173. See also "South Africa's Nuclear Weapons Program: An Annotated Chronology, 1969-1994, available at http://www.cns.miis.edu/ research/safrica/chron.htm, (3 May 2005).

79. Heitman, personal communication, 20 April 2005. See also Purkitt and Burgess,

South Africa's Weapons of Mass Destruction, 119-45.

80. De Klerk discusses this issue briefly in his autobiography. F. W. de Klerk, *The Last Trek – A New Beginning* (New York: St. Martin's Press, 1999) 273-4. For more detailed insights on the South African motivation to end its nuclear weapons program, see Purkitt and Burgess, *South Africa's Weapons of Mass Destruction*, 74, 119-45; Howlett and Simpson, "Nuclearisation and denuclearization in South Africa," 154-73; Stumpf, "South Africa's Nuclear Weapons Program: From Deterrence to Dismantlement," 4-7; and Steyn et al, *Armament and Disarmament*, 95-104.

81. According to the official findings of the International Atomic Energy Agency (IAEA) South Africa's research into weapons-related nuclear applications included the gun-type devices actually built, along with the possibilities of implosion and boosted-fission devices. See *Report by the Director General: The Agency's Verification Activities in South Africa* (Vienna: International Atomic Energy Agency, 1993). See also Roy E. Horton, III, *Out of (South) Africa*, 12.

82. According to industry insiders, there was some discussion in South Africa about production of a nuclear warhead for the new South African 155mm artillery, but the discussion apparently never progressed to an articulated requirement. These accounts somewhat conflict with other allegations that work actually was being done on such a device. See Peter Houghman and Steve McQuillen, *The Mini-Nuke Conspiracy: Mandela's Nuclear Nightmare* (London: Faber and Faber, 1995), 179. Still, miniaturization of nuclear weapons for use in artillery projectiles is a complex and expensive proposition and the author found no evidence of any serious South African developmental work on such a project, although this did not prevent analysts outside of South Africa from fretting about the possibility. Signe Landgren, *Embargo Disimplemented: South Africa's Military Industry* (New York: Oxford University Press, 1989), 93.

83. For details, see Purkitt and Burgess, *South Africa's Weapons of Mass Destruction, 134-9*; Batchelor and Willett, *Disarmament and Defence*, 71. A very interesting insider account of the US role in this process is offered by former Clinton Administration official Henry Sokolski, "Ending South Africa's Rocket Program: A Nonproliferation Success," Nonproliferation Policy Education Center, Washington, DC, Fall 1993, available at http://www.npec-web.org/essay/southafrica.htm, (2 May 2005).

84. See, *inter alia*, Frank B. Pabian, "South Africa's Nuclear Weapons Program: Lessons for Nonproliferation Policy," *The Nonproliferation Review* 3, no. 1 (Fall 1995): 11; J. W. de Villiers, Roger Jardine and Mitchell Reiss, "Why South Africa Gave up the Bomb," *Foreign Affairs* 72, no. 5 (November/December 1993): 98-109; Howlett and Simpson, "Nuclearization and Denuclearization in South Africa"; Timothy J. Muldoon, "South Africa: A Case Study in Nuclear Proliferation and Nuclear Rollback," *National Security Studies Quarterly* 5, no. 2 (Spring 1999): 83-106; Sokolski, "Ending South Africa's Rocket Program"; Purkitt and Burgess, *South Africa's Weapons of Mass Destruction*; Scott D. Sagan, "Why do States Build Nuclear Weapons? Three Models in Search of a Bomb," *International Security* 21, no. 3 (Winter 1996/97): 69-71; Stumpf, "South Africa's Nuclear Weapons Program," Bradley A. Thayer, "The Causes of Nuclear Nonproliferation and the Utility of the Nuclear Nonproliferation Regime, *Security Studies* 4, no. 3 (Spring 1995): 449-94.

85. A. R. Newby-Fraser, *Chain Reaction* (Pretoria: Atomic Energy Board, 1979), 17.

86. Purkitt and Burgess, *South Africa's Weapons of Mass Destruction*, 30-41.

87. For details, see Renfrew Christie, "South Africa's Nuclear History" (paper presented at the Nuclear History Program Fourth Annual Conference, Nice, France, 23-27

June 1993). Christie himself was tried and imprisoned by the South African state during the *apartheid* era for violation of the disclosure laws.

88. For details, see Donald B. Sole, "The South African Nuclear Case in the Light of Recent Revelations." (Paper presented at the conference, New Horizons in Arms Control and Verification, John G. Tower Center for Policy Studies, Southern Methodist University, Dallas, Texas, October 1993). See also Purkitt and Burgess, *South Africa's Weapons of Mass Destruction*, 34-7.

89. Horton, *Out of (South) Africa*, 4-5.

90. A new state entity, the Uranium Enrichment Corporation (UCOR), was created to manage the Valindaba enrichment facility. UCOR merged with South Africa's Atomic Energy Corporation in 1982. For details, see David Albright, "South Africa and the Affordable Bomb," *The Bulletin of Atomic Scientists* (July/August 1994): 37-47; Steyn *et al*, *Armament and Disarmament*, 31-4; and Stumpf, "South Africa's Nuclear Weapons Program," 4.

91. Purkitt and Burgess, *South Africa's Weapons of Mass Destruction*, 39.

92. Ibid., 41.

93. Ibid., 40-2.

94. Steyn *et al*, *Armament and Disarmament*, 40-4; Horton, *Out of (South) Africa*, 6-7; Purkitt and Burgess, *South Africa's Weapons of Mass Destruction*, 43-4.

95. In recent years, the South African media has published details of secret South African government internal consultations in the mid-1970s about the possibilities of purchasing Israeli *Jericho* missiles with nuclear warheads that would have provided the South Africans sufficient range to reach the capital cities of several hostile neighboring states. See Michael Schmidt, "Proof of SA Nuclear Plan," (Johannesburg) *Sunday Times*, 12 October 2003, 5.

96. Steyn *et al*, *Armament and Disarmament*, 76-9; Horton, *Out of (South) Africa*, 9-10.

97. Steyn *et al*, *Armament and Disarmament, 92*. For the darker suspicions, see also John Adams, *The Unnatural Alliance*, London: Quartet Books, 1984.

98. Horton, *Out of (South) Africa*, 12-4.

99. Steyn *et al*, *Armament and Disarmament*, 76. Heitman, personal communication, 20 April 2005.

100. Ibid.

101. Ibid.

102. Steyn et al, *Armament and Disarmament*, 79; Purkitt and Burgess, *South Africa's Weapons of Mass Destruction*, 73.

103. See inter alia, Adams, *The Unnatural Alliance*; Beit-Hallahmi, *The Israeli Connection*; Seymour M. Hersh, *The Samson Option* (New York: Random House, 1991); Benjamin M. Joseph, *Beseiged Bedfellows: Israel and the Land of Apartheid* (New York: Greenwood Press, 1998).

104. Stephen P. Cohen, The Idea of Pakistan (Washington, DC: Brookings Institute Press, 2004), 80.

105. Horton, *Out of (South) Africa*, 25-7.

106. Purkitt and Burgess, *South Africa's Weapons of Mass Destruction*, 15.

107. See, for example, Robert S. Jaster, "Pretoria's Nuclear Diplomacy," *CSIS Africa Notes* Number 81 (22 January 1988); Steven Laufer and Arthur Gavshon, "The Real Reason for South Africa's Nukes," (Johannesburg) *Weekly Mail & Guardian* (1 April-26 March 1993), 3; Purkitt and Burgess, *South Africa's Weapons of Mass Destruction*, 79-80.

108. Helmoed-Römer Heitman, *The South African War Machine* (Novato, CA: Pre-

sidio Press, 1985), 129.

109. This designation stands for Gun, Canada, 45 calibres. "45 calibres" indicates the length of barrel as 45 times the diameter of the bore. The standard NATO gun of the time (the US M-109) had a barrel length of only 39 calibres, significantly shorter than Bull's GC-45. Both the South African guns and the NATO standard designs have now progressed to barrel lengths of 52 calibres.

110. Landgren, *Embargo Disimplemented*, 88-95.

111. Ibid. Despite the purported US connections, the US Customs and Senate investigation of the affair in 1978 found the Space Research Corporation-Quebec (SRC-Q) guilty of violating the international arms embargo against South Africa. Bull's enterprise was bankrupted and Bull himself ultimately was tried, convicted and briefly imprisoned in the US for his role, an outcome that left him understandably embittered.

112. Landgren, *Embargo Disimplemented*, 89-92; James P. McWilliams, *Armscor: South Africa's Arms Merchant* (London: Brasseys, 1989), 72.

113. McWilliams, *Armscor*, 72.

114. Ibid., 89.

115. Bull's Space Research Corporation-Quebec was bankrupted by the scandal over breaches of the South African embargo and he apparently sold the GC-45 design to the Austrian Voest Alpine/Noricum firm, which modified it for large-scale production. The Austrian gun is designated the GHN-45. This version subsequently was purchased in quantity by Iraq. See http://www.globalsecurity.org/military/world/rsa/g5.htm, (12 March 2005).

116. Technical data derived from Denel on-line information, http://www.denel.co.za, (20 March 2005); see also Landgren, *Embargo Disimplemented*, 93-4.

117. Fred Bridgland, *The War for Africa: Twelve Months that Transformed a Continent* (Gibraltar: Ashanti Publishing Limited, 1990), 15-6.

118. Ibid., 67, 68, 87, 134, 141, 179, 182, 221, 222. Bridgland says that the South Africans deployed only three G-6s in this operation and that they were "tool room prototypes" still undergoing South African Army testing, although they apparently performed very satisfactorily in the operation.

119. For details and brief evaluation, see http://www.globalsecurity.org/military/world/rsa/g5-52.htm, (20 March 2005) and http://www.globalsecurity.org/military/world/rsa/g6-52.htm, (20 March 2005).

120. See http://www.denel.co.za, (22 March 2005) for additional detail.

121. Denel's *Seeker* RPV is a capable system, but the South African Army actually has purchased the competing ATE *Vulture* design for use with its G-5 and G-6 artillery systems. Heitman, personal communication, 20 April 2005.

122. The US Army took its workhorse UH-1, added a variety of light automatic weapons and rocket armament, and designated it the UH-1C. The Rhodesians modified their French-made *Alouette* IIIs, in somewhat the same way, calling them "K-cars."

123. Landgren, *Embargo Disimplemented*, 75-7; McWilliams, *Armscor*, 44-5.

124. The empty mass of the *Rooivalk* is 5910 kilograms compared to the *Hind's* 8,500 Kg and the *Apache Longbow's* 5360 kg.

125. This engine is a trademark product of the French-owned Turbomeca firm, a world leader in turbine engines. It is standard equipment on the *Super Puma*, *Cougar* and Eurocopter EC 725-225.

126. The last of these was delivered to the South African Air Force base at Bloemspruit in March 2004. *ARMSCOR Annual Report*, 2003-2004, 28.

127. Information provided by Paul J. Gerber, director of the South African Air Force

Fighter Programme at the time, private communications, 4 June and 7 June 2004.

128. Ibid.

129. According to Botha, *South African Defence Industry*, 7, these are: the G-5 and G-6 artillery systems (to include ammunition and propulsion); turrets and gun systems for ground combat vehicles; aircraft observation and sighting systems; air-ground tactical missile systems; laser range finders and periscopes; mine-protected vehicles; missile launch warning systems and low-cost radar warning receivers; health and utilization monitoring systems for avionics; electronic fuses for artillery and aerial bombs.

130. Ibid., 7-8. These include manufacture of airframe sections and aircraft gearboxes, aircraft engine maintenance, aircraft upgrading and weapons integration (to include the systems engineering and project management) and general systems engineering, project management, upgrading and reconditioning of military equipment.

131. This is based on the author's interviews with mid-level officials in Denel, Grintek and Alvis OMC (now BAE Land Systems OMC) in mid 2003. As an example, while the author was touring the OMC plant near Johannesburg in June 2003, he was shown a prototype command vehicle being built for US special operations forces in Iraq.

132. This is illustrated in the G-6 Self-Propelled Gun. Here, by way of example, the South Africans build a system with a gun and turret by LIW (Denel), ammunition by Naschem (Denel), a chassis and propulsion system by BAE Land Systems OMC, a fire control system by ADS and (conceivably) a remotely piloted vehicle for targeting furnished by Denel or ATE, all in one seamless package.

133. These include the Alkantpan weapons test range, the Paardefontein Antenna range (Armscor), the Overberg Test Range, the Paardefontein Explosives Range (Denel) and the Gerotek test facilities (Denel).

134. Heitman, *Unpublished Manuscript*, 89-90.

135. This is based on the author's interviews with Andre Nepgen, Director of CSIR/Defencetek, 20 June 2003; Arthur J.R. Oates, Director of Business Development, Denel, 12 June 2003; Brigadier Tony Savides (South African Army, retired), at the time Director of International Marketing, Alvis-OMC (now BAE Land Systems OMC), 18 June 2003; Alan Holloway, Vice Chairman of AMD and Managing Director of ANSYS, 21 June 2004 and Paul J. Gerber, Group Manager for Business Coordination and Counter-trade, Grintek Group Limited, 21 June 2004.

136. The sales actually were mediated by General Dynamics Land Systems – Canada, OMC's North American partner. Savides interview, 18 June 2003; http://www.baeland systemsomc.co.za, (21 March 2005).

137. Ibid.

138. "Mechem," *Journal of Mine Action*, June 2000, http://www.maic.jmu.edu/journal/4.2/profiles/mechem.htm, (10 July 2004).

139. In some respects, the 76mm gun technology developed for ground vehicle applications is more innovative than the larger field artillery systems. The rest of the world has gone to ever-larger armor piercing calibers (90mm to 100mm to 115mm to 120mm to 130mm). But the South Africans wanted (and produced) a much smaller, modern caliber gun system that could penetrate modern armor. This required innovative gun design, state-of-the-art propellants and advanced penetration technology, all mastered by South African industry. A number of South African firms also market advanced, automated fire control systems. The South Africans are competing effectively on the world market in various artillery applications.

140. Botha, *South African Defence Industry*, 7.

CHAPTER 5

The Arms Industry and the New South Africa

General

The foregoing discussion accounted for the historical development of South Africa's defense related industry, noting some of its peculiar features, and depicting its unique capabilities and success in the decade following the advent of majority rule. The study highlighted the breadth of capability achieved by the industry and the degree to which armament production now is dispersed throughout the economy and across the land area of South Africa. The discussion highlighted the determination of the South African government to use its arms industry to retain selected domestic production capabilities in support of long-term strategic needs. Also noted were the government's successful efforts to use the arms industry as a mechanism for national economic development (and empowerment of formerly disadvantaged communities). But the study has refrained from lauding the defense industry as an unalloyed blessing for the country.

The *apartheid* years saw the development of a remarkable South African capacity to produce arms, the result of a concerted, self-sacrificial effort to attain military advantage over various foes, foreign and domestic, real and imagined. The ultimate objective was security for a privileged minority community. Yet despite its very real achievements, and the military capability it produced, the industry could not avert the social changes it was created to prevent. In fact, the industry's demands on the nation's resources may have contributed to the demise of the very political establishment that created it. Even today it is by no means clear that a modest continuing investment in defense or armaments is the best use of the country's scarce national resources. Nonetheless, an important story here is the apparent determination of South Africa's new leaders to "play the hand they were dealt," seeking ways of aligning an asset of the old South

Africa with the priorities of the new. An even more interesting story is the way in which South Africans achieved a national consensus on a potentially problematic issue.

Not all South Africans are supporters of a substantial defense industry. Many are philosophically opposed to arms production and sales. For others, the industry carries the stigma of government abuses in the years before majority rule, recalling an era of coercion, unaccountable governance and diversion of national resources. Few South Africans fear the prospect of an armed threat on their borders in the foreseeable future, and an enduring series of scandals, starting with arms exports to questionable clients in the mid 1990s, and continuing with underhanded financial dealings in the "Defence Packages," still troubles relations between the industry, the government and the public. Still, South Africa's attentive public and policymakers seem to have mutually examined the defense industry and, after prolonged debate, elected to wager on its future benefits rather than sustain the resentments of the past.

It is true that the threats facing South Africa at the beginning of the 21st Century did not seem to require the sophisticated, high technology military capabilities the South Africans are seeking to maintain. The country's recent external military involvements clearly have not demanded the sophisticated conventional military forces that the country still maintains, a fact that South African scholars have been quick to note. However, forecasting the future is a tricky business, and failures of judgment in anticipating "high-end threats" can have very serious consequences. History may yet commend South Africa's current leaders for the acuity of their vision.[1]

In South Africa's decisions to maintain a defense industry capable of state-of-the-art technology, critics may see nefarious dealings within a military-industrial complex indicating little more than a successful effort by a powerful industry cynically striving to maintain status and jobs, facilitated by venal civil servants. But future benefits from a robust arms industry envisioned by South Africans go well beyond the simple economics of the present. South African technology is a source of national pride. And South Africa's leaders seem determined that their country will be an international actor and a bulwark of stability in a troubled region. They apparently have concluded that these intentions require credible military instruments of power, and the autonomy to maintain key military capabilities regardless of present external circumstances.

La plus ça change

The capabilities of South Africa's defense related industries are impressive and (apparently) growing, as the South Africans adapt to a volatile world market and the escalating commercial possibilities of new security technology. The industries themselves are basking in the glow of a remarkably successful process of accommodation to the priorities and aspirations of a new, democratic South Africa, even if it is an accommodation that still is working itself out. These facts are set against a background of dramatic ongoing political and social change in South Africa. Everything is changing. Or is it?

A remarkable feature of South Africa's arms industry is its continuity with the past and its retention of enduring features despite changes in the political environment. In 2005 South Africa still had an arms industry of intermingled public and private sector enterprises connected (and subsidized, if required) by the government, a heritage of the *apartheid* era and based on a model established as far back as World War II. In a process overseen by Armscor (itself still part of the Department of Defence), the South African military still drew its armaments from the same three sources as in the era before majority rule: from-foreign suppliers, public sector producers and private sector contractors. South African arms manufacturers still manufactured a very wide assortment of military goods and services, ranging from the mundane to the exotic, maintaining the crosscutting relationships that permit rapid and effective subcontracting to produce goods carefully tailored to customer requirements, just as they did in the *apartheid* era. They displayed an abiding willingness to compress development and production cycles, leading to rapid fielding of combat equipment, reminiscent of the 1980s. All this should not be surprising. Many of South Africa's industrial leaders in 2005 were prominent in the *apartheid* era industries two decades earlier.

Still, despite the striking continuities, there were some very significant differences between the South African arms industries of 2005 and those of the *apartheid* era. All of them, including particularly the public sector enterprises, were substantially more transparent and accountable to the nation as a whole. They (and the South African government) were much more sensitive to domestic and world opinion, and did not egregiously flaunt international norms. Despite occasional accusations of arms shipments to questionable foreign clients, they did not engage in the *apartheid* era practices of wholesale theft of intellectual property rights or shipment of sophisticated weapons to rogue states. But by far the most significant divergence from the past was the new kind of partnership between the government and the arms industry. It was much more than just government supervision. It was an interactive process of mutual support.

In official policy documents as well as in practice, the South African government emphasized its role in coordinating the efforts of public and private sector arms industries in support of the South African National Defence Force. This was articulated in the published intention to preserve certain "strategic capabilities." While this might appear to be a striking continuity with the practices of the past, it also is something of a fig leaf. Unlike their *apartheid* era predecessors, whose main focus was the needs of the Defence Force, both the arms industries and their government partners now were very externally oriented, increasingly emphasizing arms exports. South African official policy documents acknowledged that the new South Africa cannot achieve – and will not pursue – the self-sufficiency in arms sought by *apartheid* era leaders. That quest had been replaced by a search for market share and competitiveness in niche roles, concerned as much with foreign sales as with the needs of South Africa's military.

South Africa's arms producers no longer faced the restrictions of UN sanctions, and their external orientation had been facilitated by extensive partnerships with foreign arms producers, including outright foreign ownership in whole or part of South African enterprises. Nor were the South Africans oriented solely on partnerships with the developed Western nations. They were pursuing industrial relationships and coordinated marketing in a unique axis with Brazil and India. They also clearly were interested in helping their African neighbors make more informed choices about arms acquisitions.

In the first decade of majority rule, South Africa's leaders apparently came to see in a sophisticated arms industry an opportunity to pursue two objectives: limited strategic autonomy and national economic development. However, these two goals proved to be at least somewhat mutually exclusive. South Africa's defense related enterprises increasingly were connected to multinational defense industries, a trend likely to increase rather than decrease. The greater the degree of South African partnership with foreign arms producers, the better access it had both to state-of-the-art technology and to the international arms market, circumstances that were economically desirable. Yet these same connections also brought a degree of foreign influence – even control – over South Africa's industry that probably will limit the country's options in the future. Perhaps a gradual loss of strategic autonomy is the inevitable price of increasing integration into the global economy. Ironically, a failure to pursue that integration probably would preclude the access to evolving technology and economic benefits South Africa's leaders seem to deem essential to their country's security.

Change and the Future

An important advantage that South Africa enjoyed over its Cold War opponents was the sophistication of its human resources. South African engineers and technicians were (and are) "world class." For the past thirty years, as a group they may have been more capable than any analogous community of quickly designing materiel specifically suited to local military requirements. It is appropriate to ask whether or not this will be a continuing phenomenon. In other words, is the current nature of the South African arms industry a transitory historical anomaly or does it reflect an enduring organizational culture? Posed another way, will the industry's institutional norms survive its current members? A number of variables are at play here, but two issues probably are particularly important bellwether indicators of future trends: one is the degree to which the government proves willing to fund research and development, the other is the way in which the defense industry accommodates the national priority of cultural diversity in the workplace.

One key advantage enjoyed by the arms industry in the *apartheid* era was the funding allocated to unconstrained, applied research and development. At its height, this funding encompassed up to ten percent of the acquisition budget and was available to innovative teams of scientists and engineers in the public and private sectors. Insiders cite this funding as a key contributor to South Africa's impressive achievements in applied military technology. Such funding, however, ended with the advent of majority rule, and has not been replaced from any other source. In recent years, profits on commercial sales have not provided the margin from which any equivalent R&D funding can be drawn.[2] The absence of this support suggests that over the intermediate term, South Africa's arms industry will not display the innovativeness of earlier decades.

The lack of R&D funding is keenly regretted by veterans of the apartheid era arms industry, but the issue of cultural (or ethnic) diversity touches a particularly sensitive chord and its impact over the long term is harder to evaluate objectively. The organizational culture of the defense industries in the earlier era reflected the peculiarities of its white employees: a population with ties to rural agricultural communities, where mechanical skills and technical adaptation were a critical part of life, firearms were prevalent and compulsory military service a "given." The educated elite contained a high proportion of individuals grounded in engineering and the "hard" sciences. The industry itself manifested an engineering "youth culture" in which large numbers of young engineers were provided considerable latitude for creative thinking and innovation.[3] Senior political and industry leaders, scientists and engineers were drawn from the same limited population, facilitating communication and problem solving. Together, these features contributed to a remarkable ability of the South African defense industry to focus on military requirements and to compress research, development, test and production cycles, supported by the South African capacity to adapt technology from other sources to local military needs. The South Africans repeatedly progressed from an identified military requirement to a prototype design to system fielding in a much shorter time than any other defense industry familiar to the author. However, the variables that produced this capacity are changing in the new South Africa.

The infrastructure (along with many of the original personnel) of the *apartheid* era still is in place. But South Africa now celebrates its multi-cultural heritage and is making strenuous efforts to empower formerly disadvantaged communities. Much of this effort consists of affirmative action programs that are producing a multicultural workforce and managerial elite. The programs have been controversial; leading to charges that personnel actions no longer favor the best-qualified candidates. Whatever the merits of the charges, many qualified young white scientists and engineers are seeking opportunity outside South Africa. At the same time, many non-white South Africans still view the pace of change as far too slow.[4] How the South African government handles "affirmative action" will have a very tangible long-term impact on South Africa's economy as a whole, including its defense-related industries. Such decisions and choices have unpredictable consequences.

The result of affirmative action to date is a steady if slow increase in the ethnic diversity of managers, scientists and technicians in the industry as a whole, including the defense-related sector. Within a generation, South Africa's industrial leaders will come from an urban culture. They will lack the frontier ethos and the military experience of their predecessors. While the increasing diversity doubtless will bring benefits of its own, it remains to be seen if contemporary organizational values will survive the present cohort and if today's advantages will characterize tomorrow's industry.

Lessons from the South African Experience

The formation of an arms industry in South Africa was in many ways a remarkable achievement, but it is important to acknowledge that the industry was not created out of whole cloth. It drew its features and capabilities from a developed, diversified economy, comfortable with heavy industry and already producing militarily relevant goods like explosives and specialized vehicles. The country itself was relatively prosperous and generously endowed with natural resources. Yet even with this substantial base, building a capability to manufacture a comprehensive range of state-of-the-art military materiel was painfully expensive and took a quarter of a century of protracted effort. It probably seemed the only viable alternative to the governing elite at the time, but in retrospect could not indefinitely safeguard their privileged position, even if it bought just enough time to outlast the Cold War. Whatever advantages the arms industry now brings to South Africa, the actions of the *apartheid* state are hardly a recommended course of action for other developing countries. Given the sophistication of modern military technology, it may not be feasible at all, but even if it were a realistic option, establishment of such an industry surely is an alternative to be embraced only with the greatest caution.

The more interesting question is the disposition of an expensive national security resource after a significant political transition that obviates much of its original purpose. This was South Africa's dilemma in 1994. To be sure, the country's subsequent strategic choices reflected the peculiarities of leadership personalities at the time and local culture, but the choices also were the result of agreements between various political stakeholders that achieved consensus through excruciating processes of consultation. On security issues, South Africa since 1994 can take credit for broader consultation, greater accountability and more substantial transparency in its security-related decisions than most countries. Very much in contrast to the example of the *apartheid* years, South Africa's post 1994 national debate about security issues set a high standard for the rest of the world and could profitably be emulated elsewhere.

One of the most striking features of the recent evolution in South Africa's defense-related industries is the degree to which industry leaders and senior government policymakers have come to concur on national priorities in regard to the industries. These leaders certainly do not share the ethnic, cultural and ideological homogeneity of their *apartheid* era predecessors. But they have reached a remarkable agreement on three inter-related issues:

- South Africa requires the strategic autonomy provided by high-technology industry even if that capability is not economically self-sustaining.

- It is appropriate to use arms acquisitions as a mechanism to achieve a variety of key national priorities, including economic development and economic empowerment of disadvantaged communities.

- These ideas can be sold to the attentive public by relating the desired strategic capabilities to a broad notion of "security" accepted by an amorphous supporting coalition of academics and other civil society actors.

The results of this consensus so far have been the preservation of a capable industry and the promise of substantial future economic benefits. The South African example suggests that substantial cultural and ideological differences need not preclude agreement on basic themes of national security if the consultation is sufficiently broad.

The current relationship between the South African government and industry was by no means foreordained, and the roles of a number of individuals and groups were crucial to the trajectory that government-industry relations actually took. After 1994, the broader South African national leadership was at best ambivalent about maintaining an arms industry. Still, the industry survived the transition. The contribution of ANC intellectuals was important: these argued for a new defense establishment with an important role for sophisticated arms production capabilities. The new civilian leadership of Defence – particularly Joe Modise and Ronnie Kasrils – sustained the argument through national debate, but key actors from the "old" South Africa in the military and arms industry also played indispensable roles in facilitating the transition. Individuals made a difference.

Whether or not investment in defense industries is the best use of national resources, the South African use of offsets in arms acquisitions has proven a clever way of forcing foreign investment in South African industry, and an ingenious way of mandating technology transfer. It calls attention to a government willingness to use all its opportunities and explore innovative new options to promote national economic development. Because of the increasing diversification of the arms industry, and flexibility of the counter-trade agreements, new technology can be expected to benefit industrial production in sectors other than defense. Offset agreements may not be a solution to every country's arms acquisition dilemmas, but the South Africans have proven that they can be significant benefactors to an economy as a whole and to defense industries in particular.

South Africa's arms industry could not rise to its full potential until it had relatively unrestricted access to the technology of the wider world through associations with European and North American arms producers. This it has sought through vigorous pursuit of foreign partnerships since the mid 1990s and through government assistance – manifested in the novel experience of being courted by foreign producers eager to sell "Defence Package" arms in South Africa. For any country seeking to maintain state-of-the-art armament production capabilities, there really is no substitute for access to the broader currents of global technology. And even with the foreign partnerships, South Africa's success in marketing its arms has depended on an ability to discern and exploit niche roles.

Is the South African defense industry likely to be a significant global actor in the long run? It is too early to be sure. Conceivably, the country will fall victim to the vicissitudes of its own internal contradictions or the trauma of its region, though such an outcome now seems remote. More likely, South Africa will continue to exert a growing influence in world armaments markets in concert with European and North American partners and possibly in a remarkable Brazil-India-South Africa axis. But whatever its ultimate economic trajectory, South Africa's potential is significant. The power of its recent social transformation should not be underestimated and the significance of its technological capabilities should not be dismissed.

Notes

1. A historical analogy may be illustrative: had the attentive publics of the United States and United Kingdom been queried in 1932 as to the kind of military establishment most appropriate for their respective countries, they almost certainly would not have endorsed the large, complex forces their countries required to confront Axis powers a mere decade later.

2. This insight was explored in the author's interviews with Reginald Ngwenya, Director of AMD, 21 June 2004, Alan Holloway, Vice Chairman of AMD and Managing Director of ANSYS, 21 June 2004; Paul J. Gerber, Group Manager for Business Coordination and Counter-trade, Grintek Group Limited, 21 June 2004; Johan van Rensburg, commercial specialist with US Commercial Service, Johannesburg, 23 June 2004; and Helmoed-Römer Heitman, South African correspondent for *Jane's Defence Weekly*, 23 June 2004.

3. Ibid.

4. This is not only true of the non-white community. In 2004, one prominent (white) South African businessman argued in a media forum that "the children of today's captains of commerce and industry have no future in South Africa unless in our second decade of democracy, white sharing is at least equal to black forgiveness in the first." Mark Lambert (CEO, Massmart), "Whites must offer fair share," (Johannesburg) *Sunday Times Business Times*, 7 March 2004, 3.

Appendix A

Defense-Related Industries in the Public Sector

Armscor

Since its establishment in the 1960s, Armscor's key role has been to manage materiel acquisition for the South African military establishment. Between 1968 and 1992, it was the major corporate actor in South Africa's armaments industry, simultaneously a part of the Department of Defence and a large, diverse munitions-producing conglomerate in its own right. However, during the transition to majority rule (1989-1994) it underwent considerable structural change, the most significant part of which was the divestment of most of its materiel production roles. These were transferred in 1992 to a new public sector corporate entity, Denel.[1] Unlike Denel, Armscor remained a part of the South African Department of Defence after the 1992 divestment. Its new role was more explicitly to supervise defence acquisition, maintenance and materiel disposal.[2] Its remaining subsidiaries now were focused on materiel management, consultation, evaluation, testing and research.[3]

In 2005, Armscor was organized into seven main groupings, each headed by a general manager. The groupings were: Acquisition, Corporate Services, Human Resources, Quality/IT/Infrastructure, Finance, Dockyard and Armscor Business (Pty) Ltd.[4] Of these, the most relevant to the defense-related industries in general were the subsidiary enterprises of the Acquisition and Armscor Business (Pty) groupings.

The detailed activities of the Acquisition grouping were closely held and not publicly aired, but they generally included management of acquisition, procurement, product systems and technology transfer for the Department of Defence and other public sector consumers, a role made much more complicated by the responsibility to oversee the Defence Packages.[5]

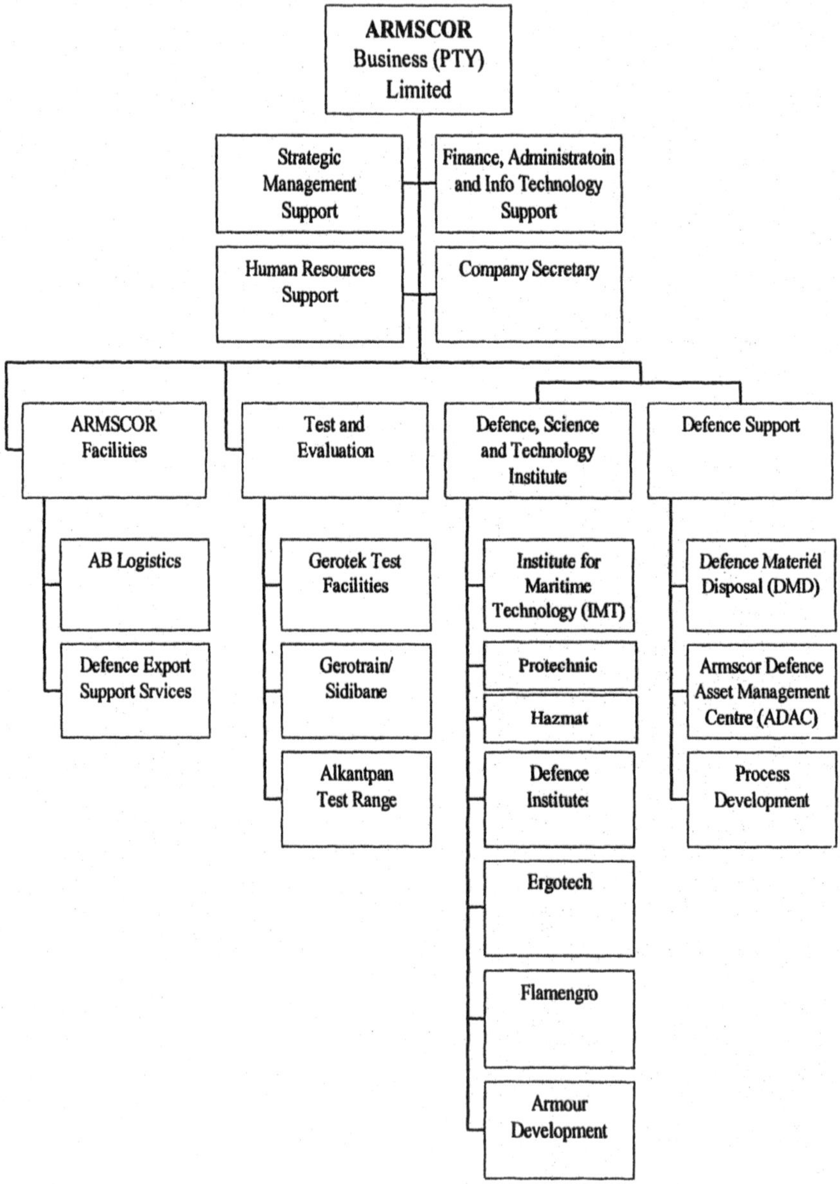

Figure A.1. Armscor Business Subsidiaries, 2004

Armscor Business (Pty) Ltd was set up in 2002 to manage what the organization calls its "strategic facilities." In 2004 these consisted of fourteen subsidiaries that offered a broad range of reimbursable services to the defense industry at large and to foreign clients. They were categorized in four distinct subgroups: Armscor Facilities; Test and Evaluation; Defence, Science and Technology Institute, and Defence Support. The individual subsidiaries within each are briefly described below.

Armscor Facilities Subgroup

Armscor Business Logistics. As its name suggests, "AB Logistics" offers logistic support services, which it provides to Armscor itself, the National Defence Force and other consumers within South Africa's armaments industry. This includes customs clearing services along with warehousing, packing and shipping by ground, sea and air.

Armscor Logistics Travel AB. This entity serves as the official 24-hour travel service and agency for Armscor and the South African Department of Defence.[6]

Test and Evaluation Subgroup

Gerotek is South Africa's best-known vehicle test facility.[7] Located in a rural area northwest of Tshwane (Pretoria), it was originally developed for military vehicle development, a role it still plays. Armscor also now offers this sophisticated facility to customers in the public and private sectors for testing of a wide range of equipment, including ergonomic material and aircraft components. Gerotek also supervises the Paardefontein Antenna Test Range (north of Tshwane) used primarily by the South African military for development of radio and microwave equipment. Gerotek migrated from Armscor to Denel in 1992, then ultimately back to Armscor.[8]

Gerotrain/Sidibane Division. Gerotrain and Sidibane are spin-off activities of the Gerotek Test Facility. Gerotrain was set up initially to provide training for military vehicle drivers, and now provides public and private sector driver training facilities, expertise and programs, including driver-training courses for the South African public. Sidibane consists mainly of a restaurant and conference center necessitated by the remoteness of the Gerotek Test Facility.[9]

Alkantpan Test Range is an all-purpose ballistic test facility located in a remote rural area near Copperton in Northern Cape Province. (The closest town with a commercial airport is Kimberley, at a distance of about 300 kilometers.) Alkantpan currently is used primarily for testing artillery systems. As South Africa's only range for heavy-caliber artillery research, it is a sophisticated center for evaluation of projectiles and explosive ordnance. It offers instrumentation at observation sites for ranges up to sixty-seven kilometers. Alkantpan's facilities and personnel are available to the South African military, defense industry and international clients interesting in testing artillery applications.[10] In 2004, Armscor named Singapore as the foreign client that had made greatest use of this range.[11]

Defence, Science and Technology Institute (DSTI)

The Institute for Maritime Technology (IMT). Armscor bills IMT as the South African Navy's "scientific support service,"[12] and also describes it as a "multi-disciplinary company specializing in defence research, development, testing and evaluating maritime systems."[13] Located in Simon's Town in the Western Cape, it offers research, development and consulting services in a wide range of surface and subsurface naval applications, including weaponry, communications and intelligence issues and equipment. In 2004 it had two subordinate business units, one oriented on naval special forces and one on customers external to the South African military establishment.[14]

Protechnik Laboratories (PTY) Ltd, located near Tshwane (Pretoria), is a company that researches defenses against chemical and biological weapons for the South African military. It also provides technical expertise on the Chemical Weapons Convention to South Africa's Council for the Non-Proliferation of Weapons of Mass Destruction.[15] In 2004, this enterprise maintained seven separate laboratories for chemical and biological research relating to synthesis, detection, materiel evaluation, carbon, filter, analysis and chromatography.[16]

Hazmat Protective Systems, located in the Tshwane area, is a manufacturing entity for respiratory protective equipment, with emphasis on defensive chemical and biological warfare. Its filters and protective cartridges are widely exported.[17] It bills itself as the only manufacturer of impregnated carbon in South Africa.[18]

The Defence Institute (located in the Armscor complex in the Tshwane area) is Armscor's in-house "think tank" on issues of security policy, military strategy and technology. Its interests extend to the identification of military materiel requirements, defense acquisition, force preparation, force deployment and decommissioning of military materiel. It offers studies and consulting services on a wide range of issues and applications, and provides multi-disciplinary approaches that overlap scientific, technical, policy and professional military expertise. The Defence Institute provides these services both to the South African National Defence Force and to other clients.[19] The institute as a whole offers "decision support methodologies and analytical tools, including modeling and simulation . . . research based on military trends and lessons learned over the full conflict spectrum . . . the use of multimedia tools . . . [and] research staff and facilitators."[20] In 2003, it maintained three separate centers:

- The Decision Support Centre concerned with "current and future defence concepts and capabilities [and] development of management concepts and capabilities."[21]

- The Acquisition Support Centre oriented on "engineering support during . . . acquisition . . . translating using requirements into system specification and project studies."[22]

- The Operations Support Centre that provided "systems management support" to users of "operational materiel."[23]

Ergonomics Technologies (Ergotech) bills itself as a "leading ergonomics consultancy."[24] It is located in the Tshwane area and services an international clientele in military and civilian health and safety applications, providing ergonomics research in "anthropometry, biomechanics and human functional performance, the specification and design of human machine systems and the evaluation of human stressors . . . and specialized health and safety issues." It also maintains an anthropometric database for the South African National Defence Force. One of its studies in 2004 involved a sophisticated evaluation of backpack technology, including measurements of "biomechanical posture, muscle activity and skin surface pressure." This research has direct and obvious application to design of human load bearing equipment for military usage.[25]

Fluid and Mechanical Engineering Group (Flamengro). Armscor describes Flamengro's product as "specialist engineering services."[26] This organization in located in the Tshwane area and maintains particular expertise in "signals analysis, structural analysis, high strain rate deformation calculations . . . and computational fluid (CFD) calculations and research." It employs "fluid and structural computer simulation software . . . to simulate complex fluid flows and to compute the stresses in and deformation of structures."[27]

Armour Development Group. This subsidiary focuses on solutions to problems of military armor protection. These range from individual body armor for soldiers and police to sophisticated main battle tank defenses, including ceramic, laminate and reactive armor designs. The group offers studies and solutions on issues of armor protection to South African and foreign clients. Its interests range from armor development to vehicle hull design to armor bonding applications, to specialized countermine protection. It also tests armor designs and applications. (In 2004, it was testing prototype armor for main battle tanks.) Its consulting services can provide specific analysis of customer requirements and advice on retrofitting of armor.[28]

Defence Support

Defence Materiel Disposal (DMD). This is the Armscor subsidiary with the mandate to dispose of excess South African military materiel on behalf of the state, typically through sales to foreign consumers. In the 2003-2004 fiscal year, DMD sold R274 million worth of excess equipment.[29] Its stock for disposal includes a wide range of munitions and ground, air and naval equipment. In 2004, DMD maintained warehouses in seven different locations (ranging from the Tshwane area to Bloemfontein to Cape Town) to accommodate its surplus stocks prior to sales.[30]

Armscor Defence Asset Management Centre (ADAC). ADAC provides a range of administrative functions for Armscor, the South African Department of Defence and Armscor clients. These services include custodianship of Departmental intellectual property, registration and management of Departmental assets, stock control (on programs and weapons systems) and identification/management of redundant, obsolete or unserviceable stock. It also is involved in marking, packaging and codification of materiel (particularly ammunition). Among other services, it offers consulting to external clients on these issues.[31]

Defence Export Support Services (DESS). DESS is an Armscor entity established in 2004 to manage Armscor's responsibilities within the South African Defence Export Support Organization (SADESO), itself a 2002 initiative that formalized the collaboration between the Department of Defence, Armscor and the South African arms industry to promote arms exports. In other words, DESS is Armscor's contribution to this key South African government initiative to emphasize arms sales to foreign consumers. Within DESS is a Technology Exploitation Center specifically designed to identify promising new applications and market opportunities.[32]

Denel

The largest corporate actor in South Africa's defence-related industries is Denel, a public sector entity that came into existence in 1992 when Armscor largely divested itself of its production enterprises.[33] In 2004, Denel was organized into seven groupings: Denel Aerospace, Denel Land Systems, Finance, Commercial, Information Technology and Procurement, Marketing, and Human Relations Training and Development.[34] Denel's arms production enterprises fell within two general groupings: Denel Aerospace and Denel Land Systems. The enterprises in these two groupings underwent considerable restructuring between 2002 and 2004, and the durability of the existing configuration was still unclear in 2005. Denel's arms production (and testing) assets in the two groupings (as of early 2005) are briefly discussed below.

Denel Aerospace

Denel Aerospace consisted of seven subordinate divisions, one of which (Denel Personnel Solutions) provided human resources support to the group.[35] The remaining six were directly related to aerospace materiel research, testing, refurbishment, assembly or production. They included:

- *Aircraft Logistics.* This relatively new entity was established in 2003-2004 by splitting Denel Aviation (itself the successor to Armscor's old Atlas Aircraft Corporation) into two parts. Aircraft Logistics is located at the Johannesburg International Airport. Its specialty is development, assembly, production and support of fixed and rotary wing aircraft.[36] It still provides technology and maintenance support to the South African Air Force. In the past (as Atlas Aircraft), it was involved in programs to upgrade the South African *Mirage* 3 fighters into the significantly more capable *Cheetah* configuration. It developed the versatile *Oryx* medium lift helicopter from the (French-designed) *Puma.* It also developed (and produced) the sophisticated and capable *Rooivalk* combat helicopter. It is a significant direct beneficiary of the South African Defence Packages. Having been part of the design team for the South African version of the *Gripen* fighters, it will build *Gripen* components, including pylons both for the South African craft and for the Swedish export version. Denel Aircraft Logistics also will final assemble, equip and test the BAE *Hawk* lead-in fighter-trainers and will build most of the South Africa's new fleet of Italian-designed Agusta A-109 Light Utility Helicopters.[37]

- *Airframe Manufacturing*. This, too, is a new entity – the other half of the older Denel Aviation that was split apart in 2003-2004. It also is based at the Johannesburg International Airport. This enterprise is chartered to produce "high-volume, low profit margin" aircraft components for both Denel and other clients. Its role includes production of aircraft sub systems and assembly of airframes. Among its assets are a fabrication plant, assembly line, machine shops and a facility for production of composite materials.[38]

- *Denel Optronics.* This is a new entity, created in 2003-2004 by merging the South African firms Cumulus and Kenis with Denel's electro-optical enterprise Eloptro.[39] It is based in Kempton Park, near the Johannesburg International Airport.[40] Eloptro, the key parent organization to this new subsidiary, was set up as an Armscor enterprise in 1974, and in time developed sophisticated electro-optical equipment including laser designators and range finders, combat vehicle sighting systems, periscope systems and components for reconnaissance satellites.[41] It still produces a wide range of materiel, including products installed in South African military vehicles, aircraft and naval vessels and in European armaments sold around the world. The technology is state-of-the-art and South Africa has had considerable success marketing its equipment abroad.[42] Another direct beneficiary of the Defence Packages, Denel Optronics designed and will manufacture the periscope for South Africa's new submarines in collaboration with the German firm Zeiss Optronic GmbH.[43] The German Submarine Consortium has adopted the Denel periscope as standard issue in its submarines and offers it to other customers on the world market.

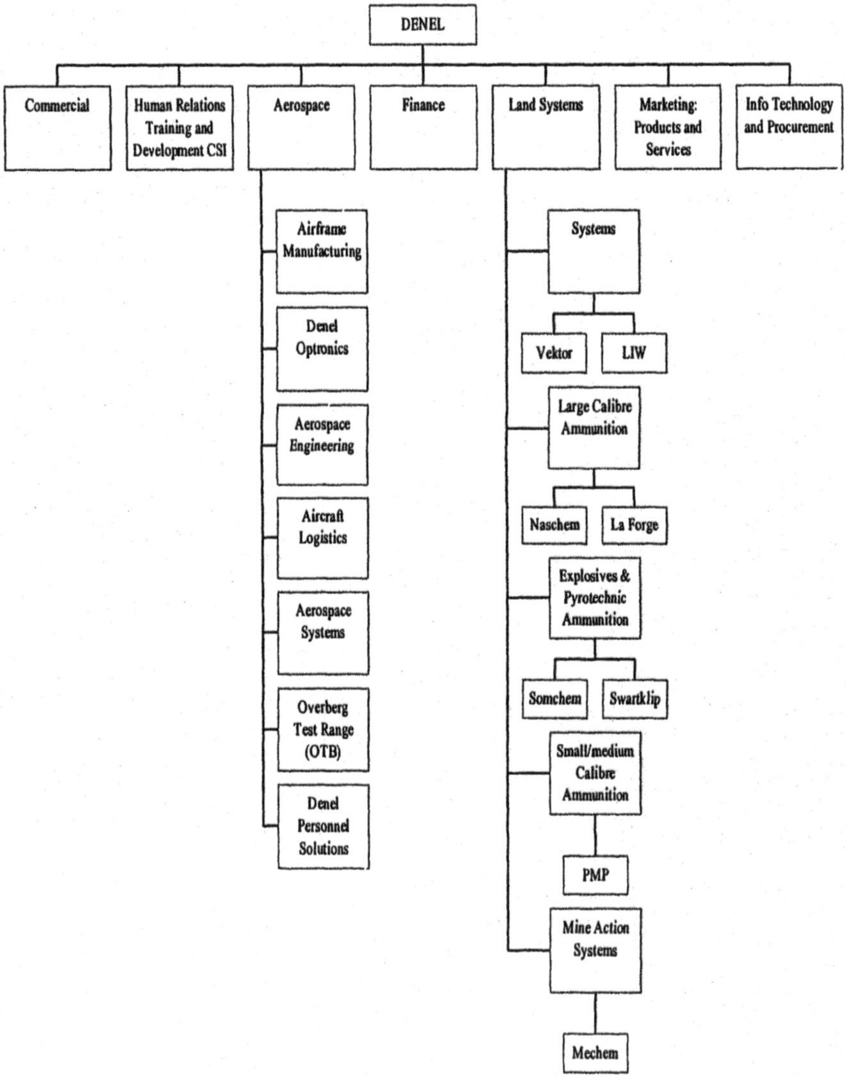

Figure A.2. Denel and its Subsidiaries, 2004

- *Aerospace Systems.* This is essentially the current version of the old Armscor subsidiary Kentron whose main expertise has traditionally been tactical missiles. The organization is based in Centurion, near Tshwane (Pretoria).[44] Its current products span the range of missiles for air-to-air, air-to-ground, ground-to-ground, ground-to-air, and anti-ship applications and reflect state-of-the-art technology in range, lethality and guidance systems.[45] In recent years, it has expanded into remotely piloted vehicles (RPVs), optical turrets for aircraft, and automated artillery command and control systems. It has had substantial success in marketing its products abroad.[46] Aerospace Systems is another direct beneficiary of the Defence Packages. Its *R-Darter* air-to-air missile currently is used by the South African Air Force and will be mounted on the new *Gripen* fighters.[47] Its *Umkhonto* ("Spear") surface-to-air missile will be installed in the Navy's new corvettes.[48]

- *Overberg Test Range (OTB).* This division, located near Bredasdorp in the Western Cape consists essentially of a sophisticated test facility for ballistic missiles and air-launched ordnance. (The major impetus for its development was the South African space program in the 1980s intended to provide a ballistic missile delivery system for South Africa's nuclear weapons.) OTB has been characterized as one of the world's best launch locations, a tribute to its geographical location, weather, and remoteness from large human populations. Co-located with a South African Air Force test range, Overberg is used for testing South African systems and also has been successfully marketed as a test and launch site to European customers.[49]

- *Aerospace Engineering.* This division particularly is concerned with design, analysis, integration and testing of aerospace equipment and subsystems. Its interests extend to weapons and aircraft stores integration, aerospace electrical systems, mission planning/debriefing, and associated software development. It also is concerned with ground testing and flight testing of aerospace equipment. In 2004, Denel's managers were contemplating the consolidation of their separate engineering departments into one division to support the entire corporation. If this occurs, the enterprise will lose its exclusive aerospace focus.

Denel Land Systems

As its current name suggests, this Denel grouping is oriented to ordnance and ground systems, billing itself somewhat grandly as "the world leader in long-range artillery systems and products."[50] In 2003, it contained eight subsidiary divisions; however, by mid 2004 these had been consolidated into five broader divisions (Systems, Large Calibre Ammunition, Explosives and Pyrotechnic Ammunition, Small/medium Calibre Ammunition and Mine Action). Each is briefly discussed below.

Systems. This division's core expertise is the design, development, assembly and support of gun systems, including a variety of military turrets.[51] The gun systems range from small arms to the largest artillery systems produced in South Africa. This division combines the operations of Vektor and LIW.

- Vektor, based in the Tshwane area, originated as the small arms production subdivision of LIW. It separated from the latter in 1997.[52] Vector currently specializes in the development and production of small arms and light crew-served weapons, and offers state-of-the art equipment in conventional and innovative new designs. Its products include pistols, assault and hunting rifles, machine guns, automatic cannons, mortars and weapons components.[53] Vektor's weapons generally receive high marks from small arms experts, and have been widely used by the South African military, but the firm has had some difficulty exporting to a world market already saturated with small arms.[54]

- LIW (Lyttleton *Ingeneurswese Werke*), also based in the Tshwane (Pretoria) area, was one of Armscor's first two subsidiaries, with an initial focus on production of small arms. It now concentrates on artillery systems, to include launchers and turrets.[55] This is the enterprise that builds the very successful G-5 towed artillery and G-6 self-propelled artillery, as well as the gun turrets for armored vehicles and ships.[56] LIW's artillery, gun system upgrades and turrets have been sold to a number of countries,[57] and in late 2003, LIW signed an agreement with the US defense firm General Dynamics Land Systems (GDLS) to jointly develop, build and market a new lightweight 105mm artillery system.[58] A prototype of this gun was tested on the US Army's new *Stryker* combat vehicle in 2004.[59] LIW is another direct beneficiary of the Defence Packages, having been selected to produce the high-tech 35mm dual purpose automatic gun systems and upgrade the 76mm main gun to be installed on the South African Navy's new corvettes.[60] LIW also has constructed experimental turrets for a new South African multi-role combat vehicle and may build the turrets and gun systems for the finalized vehicle design.[61]

Large Calibre Ammunition. This division's core expertise is the design, development, manufacture and sale of large caliber artillery ammunition, components, detonators and explosives. This division combines the operations of two older enterprises: Naschem and La Forge.[62]

- Naschem, located in Potchefstroom, North West Province, is a company with roots in the 19th century South African mining industry.[63] It now specializes in medium and large-caliber artillery ammunition and related equipment, including specialized penetrator ammunition for anti-armor applications. Other products include state-of-the-art fragmentation aerial bombs and mortar ammunition. Naschem's products have been marketed with great success in the international arms market in recent years.[64] Naschem's 105mm and 155mm ammunition was selected by the US Department of Defense for its Foreign Comparative Test (FCT) program in both Fiscal Year 2003 and 2004.[65]

- La Forge is an enterprise devoted to "forged, heat-treated and machined ammunition components . . . for artillery and naval gun systems." It also markets forgings with automotive and mining applications to civilian customers.[66]

Explosives and Pyrotechnic Ammunition. This division specializes in solid fuel (rocket) propulsion, pyrotechnics, and high explosives (for military, law enforcement and commercial applications). It combines the operations of Somchem and Swartklip.[67]

- Somchem is based in Western Cape, with its corporate offices in Sommerset West and plants near Sommerset West and Wellington.[68] This enterprise has its roots in AECI, a manufacturer of mining industry explosives created in 1924. In 1962, a subdivision of AECI took over the military propellant, munitions and explosives function to service the growing needs of the South African military. As Somchem, this entity was absorbed into Armscor in 1971 to provide propellants and explosives for the South African military.[69] It played a significant role in South Africa's space program in the 1980s, when it built the booster engines for ballistic missile prototypes. It now is the country's main producer of ammunition propellants, charges, explosives, artillery warheads and rocket and missile motors. Somchem also produces firearms propellants for the civilian market (about 11 percent of its output is "non-defense" related). In 2004, it reported that about three quarters of its products were exported.[70] The firm has had considerable success in marketing its products to overseas consumers in recent years, including a major contract to supply modular artillery charges for the British Army's new 155mm howitzers.[71]

- Swartklip is a firm with a plant near Cape Town, specializing in small arms ammunition, riot-control products, grenades and pyrotechnics. Denel bills its specialty as "high explosive products for military, law enforcement and commercial use."[72] Swartklip also produces a line of 155mm artillery projectiles configured for specialized uses such as smoke screen, illumination, chaff dispenser and leaflet dispenser. It has been particularly successful in marketing 40mm projectiles, in both low and high velocity configurations.[73]

Small/medium Calibre Ammunition. This division was formerly known as PMP (Pretoria Metal Pressings) a firm that has produced small arms ammunition since 1931, and was one of the first two enterprises absorbed into Armscor in the 1960s.[74] It specializes in small caliber military and sporting ammunition (5.56mm to 12.7mm), medium caliber ammunition (20mm to 35mm) and components. Among its other products are power cartridges, cutting charges and drill bits for the mining industry, canopy fragmentation systems for high-performance aircraft, and a new (military) long-range, anti-materiel rifle. PMP was one of the DIPP beneficiaries of the Defence Packages, in this case, attracting investment by the UK firm RO Defence (a subsidiary of BAE Systems). PMP was contracted to provide components for small arms ammunition, most of which is destined for the armed forces of the United Kingdom.[75] It is one of the world's largest ammunition manufacturers, producing for the military, law enforcement and sporting market.[76] PMP's 2004 promotional literature claimed that it exported over 70 percent of its production.[77] Among its products is a line of hunting ammunition marketed in Europe and the United States.[78]

Mine Action. This division consists essentially of the firm Mechem, an enterprise that has been very active in humanitarian demining in recent years.[79] Mechem, another Denel enterprise located in Tshwane area, originated as mine-related research unit of the Council for Scientific and Industrial Research (CSIR) in the 1960s before migrating to Armscor and, ultimately, to Denel.[80] (Mechem's research units have since been sold back to CSIR.[81]) As a subsidiary division of Denel, this firm now specializes in landmine detection, demining, demining services (including training), explosive ordnance disposal, and consulting.[82] It has successfully marketed its expertise and products, particularly to the United Nations Organization, obtaining contracts for work in Mozambique (1991-2003), Angola (1996-1998), Uganda (1996), Croatia (1997-2003), Bosnia (1998-2001), Northern Iraq (1998-2003), the Democratic Republic of the Congo (2002), Eritrea (2003) and Afghanistan (2003-present).[83] Mechem designed the *Casspir* and *Mamba* mine-protected vehicles deployed in the 1980s (now produced by BAE Land Systems OMC), and pioneered the innovative Mechem Explosive and Drug Detection System (MEDDS), a holistic mine detection system employing technical systems and dogs. Mechem also produces a 20mm anti-materiel weapon that was undergoing evaluation in 2005 by the US Department of Defense.[84]

Notes

1. See, *inter alia*, Peter Batchelor and Susan Willett, *Disarmament and Defence: Industrial Adjustment in South Africa* (London: Oxford University Press, 1998), 84-101; Gavin Cawthra, "South Africa," in *Arms Procurement Decision Making, Volume II: Chile, Greece, Malaysia, Poland, South Africa and Taiwan*, ed. Ravinder Pal Singh (Ox-

ford: Oxford University Press, 2000), 146-7; *South African White Paper on Defence-related Industries*, http://www.info.gov.za/documents/whitepapers/index.htm#1999, (26 March 2005).

2. ARMSCOR *Annual Report* 2002-2003, 18 http://www.armscor.co.za, 20 February 2004; ARMSCOR *Annual Report* 2003-2004, 1, 25-31, http://www.armscor.co.za, 26 March 2005.

3. For the most authoritative articulation of this role, see recent legislation: Armaments Corporation of South Africa, Limited Act, 2003 (Act 51 of 2003) signed into law by the State President on 20 April 2004.

4. ARMSCOR *Annual Report* 2002-2003, 18; ARMSCOR *Annual Report* 2003-2004, 24

5. ARMSCOR *Annual Report*, 2003-2004, 17-8.

6. ARMSCOR *Annual Report*, 2003-2004, 45-6.

9. ARMSCOR *Annual Report*, 2002-2003, 26-8; ARMSCOR *Annual Report*, 2003-2004, 41.

10. See http://www.gerotek.co.za, (10 March 2004) for detail.

11. ARMSCOR *Annual Report* 2002-2003, 32.

7. Http://www.armscorbusiness.com, (12 March 2004); ARMSCOR *Annual Report* 2002-2003, 32; Http://www.armscor.co.za/DefenceScienceTechnology/ListCompanies.asp, (12 March 2004).

8. ARMSCOR *Annual Report*, 2003-2004, 41.

12. Http://www.armscorbusiness.com, (15 March 2004).

13. ARMSCOR *Annual Report* 2002-2003, 31.

14. For detail, see http://www.armscor.co.za/DefenceScienceTechnology/ListCompanies.asp, (15 March 2004).

15. Http://www.armscorbusiness.com, (15 March 2004).

16. ARMSCOR *Annual Report*, 2002-2003, 30; ARMSCOR *Annual Report*, 2003-2004, 36-37. For additional data, see http://www.armscor.co.za/DefenceScienceTechnology/ListCompanies.asp, (12 March 2004).

17. Http://www.armscorbusiness.com, (15 March 2004).

18. ARMSCOR *Annual Report*, 2002-2003, 30; ARMSCOR Annual Report, 2003-2004, 37. For detail, see also http://www.hazmat.co.za, (14 March 2004).

19. ARMSCOR *Annual Report*, 2003-2004, 39; Http://www.armscorbusiness.com, (15 March 2004).

20. See http://www.armscor.co.za/DefenceScienceTechnology/ListCompanies.asp, (12 March 2004).

21. ARMSCOR *Annual Report* 2002-2003, 28.

22. Ibid.

23. Ibid.

24. Http://www.armscorbusiness.com, (12 March 2004).

25. ARMSCOR *Annual Report*, 2003-2004, 39; http://www.armscorbusiness.com, (27 April 2005).

26. ARMSCOR *Annual Report*, 2003-2004, 37.

27. Ibid.

28. Ibid., 36; http://www.armscorbusiness.com, (12 March 2004).

29. This equated to about US$ 45 million at the prevailing rate of exchange.

30. ARMSCOR *Annual Report*, 2003-2004, 42.

31. Ibid., 42- 3.

32. Ibid., 43-4.

33. See, *inter alia*, Batchelor and Willett, *Disarmament and Defence*, 84-101; Cawthra, "South Africa," 146-7; *South African White Paper on Defence-related Industries*, 1999.

34. Denel *Annual Report* 2004, 3. http://www.denel.co.za, (27 March 2005).

35. Ibid., 23.

36. Denel *Annual Report 2003*, 47-51.

37. See David Botha, *Offsetting the Costs of SA's Strategic Defence Packages* (Pretoria: Institute for Security Studies, ISS Occasional Paper 75, 2003), 6-7.

38. Denel *Annual Report*, 2004, 22.

39. Ibid.

40. For historical detail, see Signe Landgren, *Embargo Disimplemented: South Africa's Military Industry* (New York: Oxford University Press, 1989), 142-3; Batchelor and Willett, *Disarmament and Defence*, 94.

41. Batchelor and Willett, *Disarmament and Defence*, 30.

42. Denel *Annual Report*, 2003, 53-5; see also http://www.eloptro.co.za, (16 March 2004).

43. Botha, *Offsetting the Costs*, 6; and WorldSecurity-index, 2003-07-21, http://www.worldsecurity-index.com, (25 February 2005).

44. See Denel *Annual Report* 2003, 56-58 and http://www.kentron.co.za, (18 March 2004). Kentron originally coalesced as an informal group of scientists within CSIR before migrating to Armscor in the 1960s, acquiring its name and distinct function in 1968. It traditionally has been the center of South African tactical missile research and production. It played various key roles in South Africa's nuclear and space programs from the 1970s into the 1990s. Batchelor and Willett, *Disarmament and Defence*, 30. For additional historical detail, see Landgren, *Embargo Disimplemented*, 53-4, 88, 104-11.

45. See, *inter alia*, "South Africa Display Features Missiles from Kentron," Aviation Week ShowNews online http://www.aviationweek.com/shownews/02fidae/topsto17.htm, (18 March 2004) and "Surveillance Solutions," global-defence.com 2000, available at http://www.global-defence.com/2000/pages/kentron.html, (18 March 2004).

46. Among its recent successes are sales of its *Umkhonto* ("Spear") surface-to-air missile system to the Finish Navy. David Botha, *South Africa's Defence Industry, Charting a New Course?* (Pretoria: Institute for Security Studies, ISS Occasional Paper 78, 2003), 12.

47. See "Kentron's *R-Darter* becomes the SAAF's V4," *Defence System Daily*, 6 September 2000, http://defence-data/com/africa2000/pageafa15.htm, (21 March 2004).

48. See Botha, *Offsetting the Costs*, 6.

49. For details, see http://www.otb.co.za, (19 March 2004) and Denel *Annual Report* 2003, 58-9. Denel contracted with the German firm BWB in 1998 for tests of the Taurus weapons system. In late 2002, Denel announced that it was conducting discussions with Russian aerospace companies about the use of the facility for launch of low-orbit satellites. "South Africa's Denel Holding Talks with Russia on Low-orbit Satellites" Global News Wire-Asia Africa Intelligence Wire, BBC Monitoring International Reports, 23 November 2002.

50. Denel *Annual Report*, 2003, 64.

51. Denel *Annual Report*, 2004, 26.

52. LIW produced the small arms used by the South African military since the 1960s, beginning with production under license of the Belgian-designed Fabrique Na-

tionale FAL rifle, locally designated the R-1. In the 1960s LIW also obtained a Belgian (FN) license for local manufacture of the Mag-58 medium machinegun and the (Israeli designed but Belgian produced) Uzi submachinegun. It later produced a local version of the Israel *Galil* assault rifle (which South Africa calls the R-4 and R-5 series).

53. Denel *Annual Report*, 2003, 75-7 and http://www.global-defence.com/2000/pages/vektor.html, (21 March 2004). For an overview of specific products, see http://www.army-technology.com/contractors/machine_guns/vektor/, (21 March 2004).

54. See Denel *Annual Report*, 2002, 63; and Botha, *South Africa's Defence Industry*, 8.

55. Denel *Annual Report*, 2003, 67-8. For historical detail, see Landgren, *Embargo Disimplemented*, 38, 44, 88-95; and Batchelor and Willett, *Disarmament and Defence*, 30, 93, 96-7.

56. LIW typically produces this equipment in partnership with other organizations such as BAE Land Systems OMC. (The latter builds the basic vehicles and chassis for the *Rooikat* armored reconnaissance vehicle and G-6 gun.) Author's interview with Brigadier Tony Savides, South African Army, retired, Director of Overseas Marketing, Alvis-OMC (now BAE Land Systems OMC), 18 June 2003. LIW also has designed and sold new generation turrets for combat vehicles manufactured elsewhere, such as the Soviet-designed T-72 tank and the similar Indian *Arjun*, http://www.liw.co.za/faq, (25 September 2004).

57. Recent customers of the G-5 and G-6 guns have included Oman, Malaysia and the United Arab Emirates. Richard A. Bitzinger, *Towards a Brave New Arms Industry* (New York: Oxford University Press, IISS Adelphi Paper 356, 2003), 46. South Africa has conducted extensive negotiations with India over a very large proposed purchase, but in 2005 this transaction seemed to have been postponed due to issues of financing.

58. This program probably involves the innovative new LIW G7 Light Experimental Ordnance (LEO) towed gun/howitzer. See "GD teams with Denel for artillery solutions," *Military Procurement International* 13, no. 21 (1 November 2003); and http://www.dapss.com/ MPI/samples/Vol.13/issue2103/2103P-united_states2.htm, (15 December 2004). For information about the LEO gun, see Helmoed-Römer Heitman, "Denel to unveil 105mm gun with long range," *Jane's Land Forces*, 31 August 2000, available at http://www.janes.com, 22 November 2004.

59. Information provided to the author by Duncan Lang, contractor to the Office of the Secretary of Defense, 12 March 2004.

60. Botha, *Offsetting the Costs*, 6.

61. This is Project *Sepula* involving the acquisition of up to 1900 new armored fighting/troop transport vehicles to replace the aging *Ratel*. See Helmoed-Römer Heitman, "Opportunity Knocks," *Janes Defence Weekly*, 3 August 2005, 29; and Denel *Annual Report*, 2002, 52.

62. Denel *Annual Report*, 2004, 26.

63. For historical detail, see Landgren, *Embargo Disimplemented*, 95, 125-6; and Batchelor and Willett, *Disarmament and Defence*, 30, 96-7, 178.

64. Denel *Annual Report*, 2003, 71-2; see also http://sadid.co.za/SADID_7/edition7/naschem/naschem.html, (23 March 2004), and http://www.worldsecurity-index.com/details.php?id=761, (23 March 2004).

65. Scott R. Gourley, "Foreign Comparative Test Program," *Army Magazine* (February 2003) http://www.ausa.org/www/armymag.nsf, (15 September 2004).

66. Denel *Annual Report*, 2002, 53, 55.

67. Denel *Annual Report*, 2004, 26.

68. See http://www.somchem.co.za, (26 March 2004) and Denel *Annual Report* 2003, 68-71.

69. For historical detail, see Landgren, *Embargo Disimplemented*, 124-5; and Batchelor and Willett, *Disarmament and Defence*, 30, 94.

70. See http://www.somchem.co.za, (26 March 2004).

71. *Dispatch Online* (28 June 1999) http:www.dispatch.co.za/1999/06/28/ business/DENEL.htm, (21 September 2004); "155mm AS90 self-propelled howitzer," *Military Periscope Online* (1 June 2002) http://www.periscope.ucg.com/mdb-smpl/weapons/ artguns/selfprop/w0002572.shtml, (25 September 2004).

72. Denel *Annual Report*, 2003, 73.

73. See "Swartklip Products," *WorldSecurity-index.com*, http://wwwworldsecurity-index.com/details.php?id=158, (2 November 2004).

74. For historical detail, see Batchelor and Willett, *Disarmament and Defence*, 30, 37; and Landgren, *Embargo Disimplemented*, 124, 126.

75. See http://www.pmp.co.za, (27 March 2004); see also Denel *Annual Report*, 2003, 77-8.

76. Denel *Annual Report*, 2004, 26.

77. See http://www.pmp.co.za, (27 March 2004); see also Denel *Annual Report*, 2003, 77-8.

78. Ibid.

79. Denel *Annual Report*, 2004, 26.

80. For historical detail, see Batchelor and Willett, *Disarmament and Defence*, 94, 158 and http://www.mechemdemining.com, (27 March 2004).

81. "Mechem," *Journal of Mine Action*, version 4.2, June 2000 http://maic.jmu.edu/ journal/4.2/Profiles/mechem.htm, (11 November 2004).

82. Denel *Annual Report*, 2003, 78-9.

83. Http://www.mechemdemining.com, (27 March 2004).

84. This weapon is designated the NTW-20. Information provided by Duncan Lang, contractor to the Office of the US Secretary of Defense, 11 March 2004.

Appendix B

Selected Major Defense-Related Actors in the Private Sector

Grintek

Grintek Group Limited, an electro-technology conglomerate headquartered in Centurion (in the Tshwane area) probably best epitomizes the "high end" technological sophistication of the South African armaments producers in the private sector. Grintek's connection to armaments has deep roots in the *apartheid* era, and the Group itself is something of a metaphor for one path of industrial evolution in South Africa.[1]

In 2005 it was a holding company, listed on the Johannesburg Securities Exchange (JSE), whose subsidiaries employed about 1450 people in facilities distributed between Tshwane (Pretoria), Cape Town and Durban. As this is written in early 2005, about a third of Grintek's shares are owned by Kunene Brothers Holdings (a family enterprise of black South Africans). Saab AB of Sweden purchased about 22 percent at the overarching holding company level in 2004 (and in early 2005 announced an intention to purchase a up to a 70 percent share). The remainder was to be traded on the Johannesburg Securities Exchange.[2]

In 2005, Grintek Group Limited consisted of three subgroups: Grintek Telecom, Grintek TCI (Pty) Ltd (Industrial) and Grintek Defence & Technologies. Each subgroup was made up of a variety of related companies and subsidiaries. Since the early 1990s, the Group had vastly diversified its interests, now providing a wide range of private sector telecommunications applications and products ranging from electric power supply control systems to mining telemetry and communications.[3]

This industrial conglomerate traces its origins to the British communications company Racal Electronics, which established a subsidiary enterprise in Pretoria in 1963, naming it Racal SMD. By 1969, its management was in South

African hands and in 1978 it was bought by the South African firm Grinaker Holdings Limited, becoming Grinaker Electronics (Grinel). In 1981, the enterprise became part of the Anglovaal Group, one of the major South African conglomerates.[4]

In Grintek's earlier form in the *apartheid* years, the South African government identified it as the private sector "house" responsible for specialized military communications and electronic warfare systems, and much of its R&D and production at the time serviced the needs of the Defence Forces. From the 1960s onward, this resulted in a steady progression of novel military applications, some homegrown and some borrowed from abroad, adapted to South Africa's security needs.[5] Among other successes in the 1960s was the deployment of frequency hopping, high frequency tactical radio communications – a world first and a technology purchased at the time from its South African inventors by the United States and other NATO military consumers.[6]

Grintek's defence interests survived the transition to majority rule after 1994, and the group still serves as the sole source for a substantial proportion of South Africa's military requirements, particularly in the realm of "information operations" applications. These mainly fall under the purview of the Grintek Defence & Technologies subgroup, which consists of two Grintek daughter companies: Grinton (Pty) Ltd, and Grintek Ewation (Pty) Ltd. [7]

Grinton's focus in 2005 was exclusively on military/security force applications, and its products deserve some elaboration below, but the products of three other Grintek daughter companies should be mentioned here as well: Grintek Ewation, Grintek Antennas and Grintek Aviation Systems, since these firms market products that have military as well as civilian applications.

- Grintek Ewation in early 2005 was dually owned (55 percent by Grintek Group and 45 percent by the German firm Ewation GmbH, which was itself part of the European multinational EADS).[8] Its products included sophisticated systems for communications intelligence, signals direction finding, communications countermeasures and spectrum management. It also sold perimeter security equipment. Grintek Ewation offered consulting services on systems design, integration, installation and maintenance. This enterprise had contracted with EADS for joint development and international marketing of state-of-the-art signals intelligence systems under the trademark MRCM.[9]

- Grintek Antennas (Pty) Ltd, another Grintek daughter company (though not part of the Defence & Technologies Division) produced antennas and RF (microwave) systems for both the civilian and military markets.[10] The equipment was competitive in performance with the best produced in the West and had the added advantage of rapid adaptability to regional customer requirements. By 2005, Grintek had marketed its military aircraft antennas in Europe and the Middle East, consumers including Germany and Sweden.[11]

- Grintek Aviation Systems, still another Grintek daughter company not part of the Defence & Technologies division, produced and serviced avionics and air traffic management systems with both civilian and military applications. It had state-of-the-art capabilities in these roles, as illustrated by a connection to Honeywell Avionics, a US-based firm. In 2005, Grintek Aviation Systems represented Honeywell in southern Africa and managed Honeywell's regional spare parts exchange pool.[12]

Returning to Grinton (Pty) Ltd, this Grintek daughter company was dually owned in 2005 by Grintek Group (51%) and the Swedish industrial giant Saab (49%).[13] It had four subdivisions: Avitronics, specializing in microwave and electro-optic systems aimed at combat platform self-protection electronic warfare; Grintek Communications Systems, specializing in tactical communications (voice and data) and customized integrated systems; Logitronics, specializing in electronics maintenance and integrated logistic support and Grintek Integrated Defence Systems, specializing in net-centric warfare command and control systems.[14]

In 2005, Grinton's products still played a vital role in South Africa's efforts to retain key military technologies. Among the "strategically essential capabilities" identified in the 1998 *Defence Review* were technologies for command and control systems, secure communications, electronic warfare and capabilities to assure survival of combat platforms (including the technology for surveillance and threat detection and electronic countermeasures).[15] These had long been central interests of Grintek, which under present South African policy can depend on the government to subsidize the capabilities if necessary.[16]

Grintek also had become a world leader in the miniaturization of aircraft communications equipment. (The installation of Grintek's communication equipment in the cockpits of South Africa's new Agusta A-109 Light Utility Helicopters has saved so much mass and space over the original Italian designs that the aircraft can accommodate two larger than standard avionics displays, a remarkable achievement.[17]) Grintek's aircraft communications packages are manufactured with export options in mind. They are designed to accommodate virtually all currently manufactured radios, encompassing all known interfaces, facilitating the integration of different radio systems.[18]

Grintek has benefited significantly both from emerging South African government arms industry policy and from the growing partnerships with foreign defense manufacturers that sprang out of the South African "Defence Packages" announced in 1998. Grintek has been a major direct beneficiary of the Defence Packages, having been selected to provide components to each, whether corvettes, submarines, Agusta A-109 Light Utility Helicopters, BAE Hawk lead-in fighter trainers or JAS-39 *Gripen* multirole jet fighters.[19]

Grintek's connection to (the Swedish firm) Saab highlights the result of South African arms acquisition policy and its impact on local defense industries, also illustrating the success of government intervention in stimulating defense sector opportunity. The Grintek-Saab relationship reflects the novel way in which South Africa has introduced its military technology to the global market in recent years and implicitly characterizes the quality of the South African product. The relationship itself grew primarily out of the South African selection of the Swedish-designed *Gripen* jet fighter in the Defence Packages. Grintek, of course, does not produce combat aircraft. But it does produce key components.

The South African NIPPS and DIPPS protocols (described in Chapter 3) obliged prospective foreign suppliers of military materiel to establish partnerships with South African industries, and to seek out mutually advantageous commercial opportunities. Saab and BAE obtained the fixed-wing aircraft contracts in the Defence Packages, energizing their search for domestic partners. As a key local source for advanced electronic warfare materiel, Grintek had earlier attracted the attention of a Swedish company – Celsius – also a producer of advanced electronic warfare systems. Celsius subsequently purchased 49 percent of Grintek's Avitronics subsidiary. Shortly afterwards, Celsius was absorbed by Saab. The South African partners apparently impressed Saab's technical experts, resulting in the formation of an equity-based joint venture between Saab and Grintek– Grinton (Pty) Ltd – by which Saab held a 49 percent interest in four Grintek subsidiaries.[20] Another result was the initiation of personnel sharing, joint research and development and joint marketing agreements between Saab and Grintek. Perhaps more significant was Saab's subsequent decision to include Grintek's avionics and electronic warfare equipment in its most sophisticated military aircraft offered on the world market.

By 2005, Grintek and Saab were participating in a joint effort to develop and field integrated electronic warfare defenses for combat aircraft. Grintek supplied sophisticated laser warning and missile approach components as well as low-cost radar warning for helicopters and transport aircraft. Saab, for its part, offered high-end radar warning for advanced fighter aircraft, jammers and electronic countermeasure equipment. (A noteworthy feature of the Grintek missile approach warning system was that it operated in the ultraviolet spectrum rather than the commonly used infrared spectrum, significantly reducing background "noise."[21]) The whole "suite" was offered in a fully integrated single module. Since the initiation of the partnership, Saab/Grintek products have been sold to customers in Europe, the Middle East and Asia.[22]

Grintek's equipment is not limited to aerospace applications. It produces the electronic support measure (ESM) equipment installed in the German submarines that are another of the South African Defense Packages. Originally derived from Grinton's airborne systems, this equipment had been miniaturized to the point that it could be mounted on top of the submarine periscope, a notable achievement. (Competing designs by foreign firms required a separate optronics mast on the submarine.) The South African technology is sufficiently advanced that by 2003 the (German) HDW shipyards were using the

Grinton equipment as standard, and had sold boats with the South African materiel to international customers.[23]

Grinton also has marketed its aircraft communications system to the Italian helicopter manufacturer Agusta, another of the suppliers in the South African "packages." Agusta was sufficiently impressed with the South African communications equipment to offer it as a standard feature on its A-109 light utility helicopters, including those now displayed at various international air shows and sold around the world.[24] Grinton has established partnerships with foreign corporations other than those directly involved in the Defence Packages. As noted above, it markets intercept and jamming equipment in cooperation with the European conglomerate EADS. By 2003 it also had been selected by (the French-owned) Thales Avionics as the regional repair authority for the latter's avionics displays.[25]

Grintek is producing "cutting-edge" communications, electronic warfare and signals intercept equipment. More importantly, it has succeeded in marketing its products to discriminating foreign consumers. Partnerships have been an important contributor – providing some technology transfer and access to the wider defense industry community in Europe. Much of this is directly or indirectly attributable to South African acquisition policy reflected in the Defence Packages.

BAE Land Systems OMC

A discussion of significant actors in the South African private sector defense industry would be remiss without some attention to the major producers of ground vehicles and vehicular sub-assemblies. The story of this production is very complex, involving a number of historically autonomous private sector enterprises and a very long list of corporate combinations, divestments and name changes, but South Africa's military vehicle industry now has come to rest essentially in one South African subsidiary of the UK-based multinational British Aerospace Industries (BAE). The South African end of this military vehicle connection is Land Systems South Africa, as this is written in early 2005, a business unit of the UK-based defense conglomerate BAE Systems Land Systems. Land Systems South Africa is a corporate group that provides most of the military (and security) vehicle production and refurbishment in South Africa. In early 2005 it was 75 percent owned by BAE Systems and 25 percent owned by South African investors.[26]

The history of military vehicle production in South Africa provides another interesting thread in the evolution of the private sector arms industry, and is worthy of a brief overview. An important early South African ancestor was the Austral Iron and Engineering Works Ltd with roots in mining and general engineering. In 1964, the South African government gave Austral a contract to produce (under license) the French AML *Panhard* wheeled armored scout vehicle, which the South Africans renamed the *Eland*. By 1968, Austral had made South Africa self-sufficient in production of this vehicle. Widely used by the South African Army in scouting, raiding and light armor roles, the *Eland* was upgraded through seven distinct local versions before production ceased in the mid 1980s.[27]

By the early 1970s, Austral had been given a contract to produce the indigenously designed *Ratel* wheeled infantry fighting vehicle, which began to enter the South African inventory in the mid 1970s. The *Ratel*, a versatile and advanced design for its time, subsequently proved a great success in combat operations in Angola in the 1980s.[28]

In 1971, Austral merged with the Sandock Ltd shipyard to become Sandock-Austral Ltd., diversifying into vehicle research and design and vehicular components. Sandock-Austral also was given a contract in the late 1970s to produce an indigenously designed armored reconnaissance vehicle to replace the *Eland*, a sophisticated new design called the *Rooikat*. The *Rooikat* began to enter the South African Army inventory in the late 1980s. By this point, enterprises connected to Sandock-Austral also were producing several distinct lines of transport, utility and combat vehicles, with particular emphasis on wheeled carriages with extensive counter-mine protection.[29]

South Africa decided in the mid 1970s to upgrade its small fleet of (British-designed) 1950s era *Centurion* main battle tanks, calling the new model the *Olifant* ("elephant").[30] It financed the creation of a new, independent enterprise in 1977 – the Olifant Manufacturing Company (OMC) Ltd – to perform this upgrade. In 1987, the research and development facilities of Sandock-Austral amalgamated those of OMC and the new enterprise was renamed Ermetek.

In 1994, Ermetek was absorbed into the holding company Reunert Mechanical Systems (Reumech). A year later, it was combined with Sandock-Austral and several other companies to form Reumech OMC. Reumech OMC acquired the rights to military vehicles formerly produced by several other South African vehicle manufacturers, notably Truckmakers and TFM Defence. This brought in lines of military cargo trucks (Samil series) and various police and security vehicles.[31]

In 1999, Reumech OMC was bought outright by the UK-based firm Vickers plc., and renamed Vickers OMC. The (South African) investment company DGD Technologies acquired a 25% share in Vickers OMC in 2002. The remaining 75% was purchased by Alvis plc later in 2002, and the company assumed the name of Alvis OMC at the time. The Alvis connection continued until BAE purchased the firm in 2004.[32]

While the BAE and (former) Alvis and Vickers connections have provided international visibility and access to otherwise difficult foreign markets, the South African group enjoys substantial autonomy and has successfully developed its own overseas commercial connections.[33] Local people manage it. Decisions on human resources, production and marketing are made locally.

As a result of the 2004 BAE acquisition, Land Systems South Africa became the South African parent company for the vehicle enterprises. In 2005 it consisted of two more or less autonomous subsidiaries: Land Systems OMC and Land Systems Gear Ratio. The former was the South African firm primarily involved in security vehicle production, the latter a specialist manufacturer of transmissions, gear shafts and related vehicle components.[34] Land Systems OMC obviously is of key interest here, so the subsequent focus turns exclusively to this subsidiary.[35]

In its 2005 promotional literature, "OMC" described itself as South Africa's primary military vehicle facility, covering all disciplines of the military vehicle spectrum from conceptualization through design and development, manufacture and production and "in service support."[36] Its current configuration it is the result of a long and complex process of consolidation in the South African military vehicle industry stretching back to the 1960s.

In mid 2003, OMC employed a work force of about six hundred, of which about 420 employees were permanent, and the rest on temporary contract.[37] The firm's principal facilities were located among other enterprises in a vast industrial park near Benoni, a few miles east of the Johannesburg metropolitan area. The central facility was a large factory structure containing a variety of assembly lines for different vehicles. Most of these lines were devoted to the refurbishment and upgrade of vehicles produced in the past. However, the company continued to manufacture several different series of wheeled vehicles with a range of security applications.[38]

At the light end of the vehicle spectrum, OMC's products included an innovative new "rapid deployment and reconnaissance vehicle" – the *Wasp* – purchased by the South African Army. This US$4.4 million contract, signed in late 2003, was for 25 of the vehicles with ancillary equipment, delivery to begin in mid 2004. The firm's products also included the older *Mamba* and *Casspir* mine-protected vehicles originally designed by Armscor's Mechem subsidiary. OMC's busier production lines included wheeled vehicles like the Scout 2 (a jeep-like vehicle), the RG-12, a heavy, wheeled, armored carrier, and the RG-31 and 32 series, medium weight wheeled vehicles with a large variety of potential configurations and applications.[39]

OMC also owned the production rights to several other specialized systems, including the *Rooikat* armored reconnaissance vehicle, which it can produce for export when the company obtains foreign orders that comply with South Africa's export regime. (OMC builds the chassis and hull and subcontracts the turret and gun systems from Denel's LIW.) The firm manufactures the chassis for the indigenously produced G-6 self-propelled 155mm gun, although Denel markets the finished weapon, producing the gun system and the ammunition. The G-6 gun system has been sold to a number of countries since the early 1990s.[40]

In 2005 OMC claimed that it was the "main supplier and maintainer of armoured vehicles for the . . . South African Army . . . the South African Police Service and local security companies.[41] When the author toured the Alvis OMC plant in June 2003, the refurbishment assembly lines contained *Olifant* tanks, *Rooikat* armored reconnaissance vehicles, *Ratel* infantry fighting vehicles, G-6 gun chassis and several other indigenously designed vehicles under South African National Defence Force contract. Assembly lines for new vehicles included mine-protected armored personnel carriers for export.[42]

OMC divided its activity into four categories: it built some new vehicles, it upgraded (and refurbished) older vehicles, it assisted clients that had purchased its products (with repairs, field support, spares and dealer services) and it conducted research. While the majority of its business in 2005 still was provided by South African government contract, it had succeeded in marketing its vehicles and services abroad, claiming sales to over twenty countries.[43] The firm had established a North American marketing arrangement with General Dynamics Land Systems- Canada, and its vehicles had been purchased by law enforcement agencies in the US and Canada and by the US military (as described in Chapter 4).[44]

As of 2005, the bulk of OMC's exports had been the RG series of wheeled vehicles. These come in a variety of configurations and can be significantly tailored for customer requirements. They all were designed specifically against the threat of landmines. Options include different levels of landmine protection. The company has been willing to custom build, to sell small numbers of vehicles and to produce on short notice.

Its most successful single item had been the RG-12 armored personnel carrier. In mid 2003, some seven hundred were in service in eight countries, and customers included Canadian and US police organizations.[45] The company's single greatest foreign marketing success up to 2005 was the sale that year of 148 of its RG-31 series vehicles to the US Army through its North American broker. The company's various overseas sales have been facilitated by its European-based connections and by partnerships with selected other foreign industries. These included a relationship with General Dynamics Land Systems Canada (GDLS-C), established in 2002, for marketing of OMC vehicles in North America,[46] and an affiliation with TCT Blindados Brazil for access to the South American market.[47]

The company also performed research that probably is best described as research into the utility of existing technology. For instance, it did not produce its own vehicle engines, but tested and bought the best it could obtain from abroad, along with a number of other components.[48] Still, there was one domain of research in which OMC may have been "cutting edge" – the prediction of landmine effects on vehicle occupants. OMC combined sophisticated computer simulations with tests of explosive effects on actual equipment to predict outcomes. By 2004, the firm had begun to study effects of the kinds of improvised explosive devices (IEDs) encountered by the US-led coalition in Iraq. OMC applied its findings to its own vehicular products, but studied the effects on vehicles manufactured elsewhere as well, claiming a high degree of confidence in the results.[49] OMC's research highlights two aspects of current South African technology – a continuing interest in attenuating the effects of mine warfare and the ability to apply sophisticated simulation in study of tactical problems. Both are capabilities that the South Africans were seeking to exploit in 2005.

OMC is a good example of a South African industry that has found a lucrative niche market. It claims to field the world's best mine-protected vehicles, a claim hard to dispute. It is the direct descendant of companies with years of experience in building mine-defeating technology. Its vehicle designs have been extensively and successfully tested in some of the world's most mine-ridden areas. OMC also is the inheritor of a long term South African investment in wheeled combat vehicle technology, which contributes to mobility and ease of maintenance – and rapid repair after mine detonations. The wheeled chassis of vehicles like the *Rooikat* and G-6 self-propelled gun were developed especially for the African operational environment and are works of engineering genius in their own right. OMC also has the advantage of considerable flexibility to customer requirements. For all of these reasons, customers that require relatively small numbers of very robust, mine-resistant vehicles and good customer support after the sale probably will find OMC an attractive supplier, and the enterprise is likely to experience continued success in overseas sales.

Reunert

Reunert Ltd, another of the major defense-related groups, also has a long history of South African defense contracts across a wide range of fields, although most of its focus now is outside the defense arena.[50] This group has its roots in the 19th Century, as the company Reunert & Lenz (established in 1888), an early South African leader in electric lighting, power generation and transmission, and heavy machinery like locomotives and steamrollers. In the middle of the 20th Century, Reunert underwent a complex series of corporate connections and restructurings, and by the mid 1960s had become an increasingly important actor in defense-related production. (In addition to its consistent focus on electronic applications, it at one time largely had absorbed the South African military vehicle industry, but has since divested itself of these enterprises.) The Reunert Group now advertises its specialty as "electronics and low-voltage electrical engineering, supplying value-added products, systems and solutions." It still provides some goods and services to the South African National Defence Force, but much of its military materiel is exported.[51]

Reunert functions as a holding company. Its corporate headquarters are located in Sandton (Johannesburg) and its daughter companies are widely distributed across South Africa. It employs a total of just under five thousand people. Its promotional literature in 2004 listed eighteen subsidiary enterprises, only four of which are focused on defense-related goods and services.[52] Its annual report for 2003 claimed annual revenue growth of 21 percent and an increase in operating profit of 29 percent over the previous year.[53]

Reunert's defense-related daughter companies are grouped under the umbrella of "Reutech," and include the following:

- Fuchs Electronics. This company, located in Alberton (Gauteng Province), specialized in electronic fusing.[54] In 2003, it employed two hundred and sixty two people, producing state-of-the-art fuses for artillery, rocket and mortar ammunition. It reported that 90 percent of its revenue was generated by exports.[55]

- Reutech Defense Industries (RRI). This company, located in Pinetown/New Germany (KwaZulu-Natal) specialized in tactical communications systems and air-delivered weapons systems.[56] In 2003, it employed one hundred and forty four people, producing very high frequency (VHF) and high frequency (HF) communications systems and electronic/fusing and warhead ordnance components.[57]

- Reutech Radar Systems (RRS). This company, located in Stellenbosch (Western Cape), specialized in search and tracking radar systems for air defense, airspace control and air traffic management.[58] In 2003, it employed eighty-eight people. The company was partially foreign owned. While Reunert owned a controlling interest of 57 percent, 33 percent was owned by the European-based defense multinational, EADS. Kgorong Investments, a black South African investment group, owned the remaining 10 percent.[59]

- Reutech Defence Logistics (RDL). This company, located in Midrand (Gauteng Province), advertised its expertise as "logistic support and management . . . gun and fire control systems for armoured vehicles and naval guns . . . [and] supply and support of . . . internetworking equipment."[60] In 2003, RDL employed two hundred and twenty-one people.[61]

Notes

1. For organizational detail, see http://www.grintek.com, (22 November 2003).

2. Interview with Paul J. Gerber, Group Manager for Business Coordination and Counter-trade, Grintek Group Limited, 12 June 2003, and personal communication with the author, 22 April 2004.

3. Http://www.grintek.com, (22 November 2003).

4. Ibid.

5. Signe Landgren, *Embargo Disimplemented: South Africa's Military Industry* (New York: Oxford University Press, 1989), 140-1.

6. Gerber interview, 12 June 2003, and personal communication, 22 April 2004.

7. Http://www.grintek.com, (22 November 2003).

8. Gerber, personal communication, 22 April 2004.

9. The MRCM acronym stands for "monitoring, reconnaissance, and countermeasures." Gerber interview, 12 June 2003 and personal communication, 22 April 2004.

10. According to Gerber, much of the incentive for development of RF communications came from South African desire in recent years to attenuate the theft of copper and electro-optical cable used in non RF communications systems, Gerber interview, 12 June 2003.

11. Ibid.

12. Ibid.

13. Http://www.grintek.com/structure, (25 October 2003).

14. Ibid; Gerber personal communication, 22 April 2004.

15. *South African Defence Review* (Pretoria: 1 Military Training Regiment, 1998), Paragraph 10. Such technologies "satisfy a unique environment-dependent need which cannot be satisfied through procurement of standard equipment, provide an operational winning edge, or [are essential to] high priority operational requirements."

16. *Defence Review*, Paragraph 102.

17. Gerber, personal communication, 22 April 2004.

18. Ibid.

19. David Botha, *Offsetting the Costs of South Africa's Strategic Defence Packages* (Pretoria: Institute for Security Studies, ISS Occasional Paper 75, 2003), 6-7; and Gerber, personal communication, 22 April 2004.

20. Gerber interview, 12 June 2003 and personal communication, 22 April 2004. According to Gerber, Saab wanted to purchase a controlling interest at the time but was prevented from doing so by South African Government policy.

21. Gerber interview, 12 June 2003.

22. Ibid. According to Gerber, although marketed by Saab rather than Grintek, up to 70% of the electronic warfare upgrade of the German Air Force *Tornado* between 2000 and 2003 was comprised of South African components.

23. Ibid.

24. Gerber interview, 12 June 2003; see also Botha, *Offsetting the Costs*, 6-7.

25. Ibid.

26. The South African investors are DGD Technologies (2001) (Pty) Ltd. The founding shareholders in this enterprise are the KMM Investment (Pty) Ltd owned by Khapametsi Maleke and Moeletsi Mbeki and Goldenwood Investments (Pty) Ltd owned by Diliza Mji. Interview with Brigadier Tony Savides (South African Army, retired), Director of International Marketing, Alvis-OMC (now BAE Land Systems OMC), 18 June 2003; Http://www.baesystemsomc.co.za, (26 March 2005).

27. Landgren, *Embargo Disimplemented*, 82-3.

28. Ibid., 24, 84; Savides interview, 18 June 2003.

29. James P. McWilliams, *Armscor: South Africa's Arms Merchant* (London: Brasseys, 1989), 54.

30. South Africa originally purchased some 200 *Centurions* from the United Kingdom, taking delivery between 1955 and 1959. It sold many of these in the 1960s, and when it subsequently decided to upgrade its *Centurion* fleet, was obliged to purchase additional *Centurions* from various countries to have enough for its desired force strength. Landgren, *Embargo Disemplemented*, 39, 85-7.

31. Helmoed-Römer Heitman, *Unpublished Manuscript* provided to the author in June 2003, 132-3; Landgren, *Embargo Disimplemented*, 81-2, 96-7; McWilliams, *Armscor*, 31, 39-40.

32. Savides interview, 18 June 2003; http://www.baesystemsomc.co.za, (26 March 2005).

33. Ibid.

34. Http://www.baesystemsomc.co.za, (26 March 2005).

35. OMC stands for "*Olifant* Manufacturing Company."

36. Http://www.baesystemsomc.co.za, (26 March 2005).

37. Savides interview, 18 June 2003.

38. Ibid; Author's observation of the OMC facilities at Benoni, 18 June 2003.

39. Ibid.

40. Savides interview, 18 June 2003.

41. Http://www.baesystemsomc.co.za, (26 March 2005).

42. Ibid; author's observation at Benoni, 18 June 2003.

43. Major recent contracts (signed in late 2003) include the SANDF purchase the *Wasp* Rapid Deployment Reconnaissance Vehicle noted above and a US $17 million contract to upgrade the South African Army's *Olifant* main battle tanks. This upgrade, in consonance with other contractors, included new powerpacks, fire control and sighting equipment. Http://www.alvisomc.co.za, (9 September 2003) and Http://www.baesystems omc.co.za, (26 March 2005).

44. Savides interview, 18 June 2003; author's observations at Benoni, 18 June 2003.

45. Ibid; http://www.baesystemsomc.co.za, (26 March 2005). The RG-12 is a sophisticated, light armored personnel carrier designed for a driver, vehicle commander and cargo capacity of up to ten additional personnel. It is in service with the Royal Canadian Mounted Police and several US law enforcement organizations. Its optional mobile adjustable ramp system facilitates its use in emergency/rescue or anti-highjacking roles.

46. Http://www.scribe.co.za/pressers/current/19_november_2002.htm, (12 September 2004). See also the South African vehicles advertised by General Dynamics Land Systems at http://www.gdls.com/main.html, (12 September 2004).

47. Http://www.alvisomc.co.za/News/news3.htm, (12 September 2004). In 2003, the Brazilians were in the process of deciding on a new family of armored vehicles for their armed forces, and OMC clearly sought to position itself to compete for this lucrative contract.

48. Ibid.

49. Savides, private communication 12 March 2004. Savides describes this research as an OMC *forte* but acknowledges "complimentary research" by other South African enterprises.

50. Http://www.reunert.co.za/about/history.htm, (7 April 2004).

51. Ibid.

52. For an overview of corporate details, see http://www.reunert.co.za/about/ structure.htm, (7 April 2004).

53. *Reunert Annual Report – 2003* http://www.reunert.com/invester/annual_2003/, (12 March 2004).

54. Http://www.fuchs.co.za, (13 March 2004).

55. *Reunert Annual Report – 2003.*

56. For details, including this overview, see http://www.rdi.co.za, (12 July 2004).

57. *Reunert Annual Report – 2003.*

58. For details, including this overview, see http://www.rrs.co.za, (12 July 2004).

59. *Reunert Annual Report – 2003.*

60. For details, including this overview, see http://www.rdlog.co.za, (12 July 2004).

61. *Reunert Annual Report – 2003.*

Appendix C

Selected Small Defense-Related Actors in the Private Sector

Cyber Related Enterprises

South African industry embraced the information revolution very early in its history, and a variety of companies offer military-related cyber applications. Among the firms prominent in this field in 2005 were:

- *Ariel Technologies* of Rivonia, specializing in network design, management and support, training and specific access control and personnel identification solutions.[1]

- *C^2I^2 Systems* based in Rondebosch, an information technology company specializing in software for combat information and management systems.[2] The firm had sold its products to NATO consumers, including Raytheon Missile Systems in subcontracts for AWACS and naval surface missile systems. It also was one of the firms directly benefiting from the Defence Packages – providing command management software for the new Class 209 submarines.

- *Cybersim* of Menlo Park, a company specializing in simulation software with particular interests in decision-support applications for combat and peace support, and disaster management operations.[3]

- *Nanoteq* of Centurion, a company with a particular interest in information security for both military and private sector usage. Among its customers were clients within the South African National Defence Forces that purchased cryptographic equipment.[4]

- *ThoroughTec* of Umhlanga Rocks, a company with expertise in simulators for various usages, military and civilian. It has endeavored to market an armored vehicle simulator to the South African National Defence Forces.[5]

In addition to the firms noted above, several other South African companies emphasized battle management systems for command and control and weapons coordination applications. Among the most prominent of these are ANSYS Integrated Systems of Waterkloof that advertised "weapons system integration, solutions and products" including information software and hardware. Its products spanned a range of ground based and air weapons, including anti-tank missile, aerial rocket and air-to-air missile delivery systems. Customers have included major contractors for the South African National Defence Forces (preeminently Denel) and the German LFK Gmbh.[6]

Communications/Electronics Enterprises

Since the 1960s, South Africa has demonstrated considerable strength in its communications/electronics enterprises. These include not only the subsidiaries of large conglomerates like Grintek, discussed earlier, but also a number of smaller, autonomous firms. In 2005, actors in this category included the following.

- *Electromagnetic Software and Systems (EMS)* of Stellenbosch emphasized antennas, also offering a range of electro-magnetic engineering services to military and civilian customers. These services included radar cross-section prediction, radiation hazard analysis, and design, construction and placement of antennas.[7]

- *Hausler Scientific Instruments (H.S.I)*, a Jeppestown firm with a wide production repertoire, specialized in microwave applications. Its other products included components for night-vision devices, artillery bore scopes, artillery recoil systems, missile and munitions fuses, aircraft engine components, and police small arms.[8]

- *Sysdel* of Pierre van Ryneveld specialized in electronic warfare applications, which it had sold in quantity to the South African Air Force and Navy. Its products typically were produced as subsystems of electronic intelligence, electronic support measures and jammer systems. One of the direct beneficiaries of the Defence Packages, Sysdel provided EW components to the South African Navy's new corvettes. It also had developed extensive foreign partnerships, including (among others) relations with the European defense conglomerate EADS, the Swiss Telephan and the US Herley Industries.[9]

- *Tellumat* of Tokai was a firm specializing in radar and avionics systems.[10] Its services included upgrades of the air surveillance radars used by the South African Air Force, and it produced the IFF transponders designated as standard issue throughout the South African National Defence Forces. Other products included IFF crypto systems, TACAN equipment and remotely piloted vehicle data links. Another direct beneficiary of the Defence Packages, Tellumat manufactured much of the combat control system for the Navy's new submarines and had been contracted to oversee much of the local software development for the system in South Africa.[11]

Aerospace Products

In 2005 Denel Aviation was the largest aerospace production enterprise in South Africa, but it was by no means the only one. Among other prominent actors in the aerospace sector were Aerosud (noted in Chapter 4, above), located near Waterkloof Air Base in the Tshwane area, Aerospace Monitoring Systems (AMS), located in Halfway House, and Turbomeca Africa located near the Johannesburg International Airport.

- *AMS* specializes in health and usage monitoring systems (HUMS), utility computers and positioning systems. The South African Air Force, the Royal Australian Air Force and the Canadian-based NATO Flying Training Center, had purchased HUMS equipment produced by this firm.[12]

- *Turbomeca Africa* was formerly known as Denel Aviation Airmotive. Denel sold 51 percent of the enterprise to the French firm Turbomeca in 2002. This firm manufactured turbine engine and gearbox components for military and civilian clients.[13] It was another direct beneficiary of the Defence Packages, having obtained the contract for maintenance and repair of the Makila engines for the Agusta A-109 Light Utility Helicopters.[14]

Miscellaneous Specialized Products

In 2005, a significant number of South African enterprises offered specialized products in narrow fields. A small, representative sample is illustrative.

- *Althane Technologies International* of Oberholzer produced polyurethane products for military usage. This firm has sold its polyurethane road wheels to the South African Army for use on its main battle tanks, and was researching puncture-proof wheels for combat vehicles. Its polyurethane wheels had been tested for possible military usage by acquisition authorities in the United Kingdom and United States.[15]

- *Delkon* of Durbanville specialized in powertrain and powerpack applications for combat vehicles.[16] It had participated in the development of a high ambient temperature tank engine and a replacement diesel engine for the South African *Eland* armored reconnaissance vehicle.[17]

- *Diel*, a firm located in Plumstead, emphasized specialized components for artillery applications.[18] Its products included a state-of-the-art meteorological ground station and an automated forward observer system.[19]

- *Kal-Gard* of Wandsbeck specialized in the development of coatings and lubricants for the military and civilian market. It had marketed corrosion-resistant coatings for weapons and ammunition. It also manufactured preservative and lubricating oils for weapons, cleaning materials, and specialized automotive lubricants.[20]

- *MGL Milkor Marketing* of Hatfield focused on special-purpose infantry weapons.[21] Milkor developed a six-shot, hand-held 40mm grenade launcher that had been widely exported. It also developed an automatic 40mm grenade launcher for use by the South African Army, though Denel actually was slated to manufacture the item.[22]

- *MOH-9 Armour Ceramics* of Pretoria, a subsidiary of Multotec Wear Linings, produces aluminum-based armor. Its products include armored tiles for vehicular, aircraft, ship and individual soldier protection.[23]

- *Parachute Industries of South Africa (PISA)* of Verulam specialized in parachute equipment. Its products ranged from personnel and cargo parachutes to drogue chutes for jet aircraft to bomb retarders, and it marketed its products extensively to the South African military.[24]

- *Stingray Marine* of Durbanville specialized in small ships with military applications.[25] Its products included inflatable craft for special forces and twin-hull, high-speed hydrofoils.[26]

- *Three Spears Africa* of Maitland researched and manufactured webbing for civilian and military usage.[27]

- *Truvelo Armoury Division* of Lyttleton produced specialized ground force small arms.[28] Its products included sniping and long-range anti-materiel rifles that it was endeavoring to market to military and police consumers.[29]

- *Excel Information Technologies* of Lynnwood Ridge offered highly sophisticated military logistics services, including advanced cyber applications. Its services included training packages. It had sold its products to a wide range of domestic and foreign consumers, including the South African military, Armscor, Denel, the Italian-based Agusta corporation, and the Royal Air Force.[30]

Notes

1. Http://www.ariel.co.za, (6 September 2003).
2. Http://www.ccii.co.za, (10 September 2003).
3. Http://www.cybersim.co.za, (6 September 2003).
4. Http://www.nonteq.com, (7 September 2003).
5. Helmoed-Römer Heitman, *Unpublished Manuscript* provided to the author in June 2003, 147.
6. Interview with Alan Holloway, Vice Chairman of AMD and Managing Director, ANSYS, 21 June 2004; http://www.ansys.co.za, (12 July 2004).
7. Http://www.emss.co.za, (7 September 2003).
8. Heitman, *Unpublished Manuscript*, 147.
9. Ibid., 150.
10. Http://www.tellumat.com, (11 September 2003).
11. Heitman, *Unpublished Manuscript*, 150.
12. Http://www.ams.co.za, (11 September 2003)
13. Denel *Annual Report*, 2003, 46, http://www.denel.co.za, (6 May 2004); David Botha, *South Africa's Defence Industry: Charting a New Course?* (Pretoria: Institute for Security Studies, ISS Occasional Paper 78, 2003), 8, 11.
14. David Botha, *Offsetting the Costs of SA's Strategic Defence Packages* (Pretoria: Institute for Security Studies, ISS Occasional Paper 75, 2003), 7.
15. Heitman, *Unpublished Manuscript*, 144.
16. Http://www.a2africa.co.za/delkon, (5 September 2003).
17. Heitman, *Unpublished Manuscript*, 146.
18. Http://www.diel.co.za, (7 September 2003).
19. Heitman, *Unpublished Manuscript*, 146.
20. Ibid., 148.
21. Http://www.milkor.co.za, (12 September 2003).
22. Heitman, *Unpublished Manuscript*, 148.
23. Ibid.
24. Ibid., 149.
25. Http://www.stingramarine.com, (6 September 2003).
26. Heitman, *Unpublished Manuscript*, 150.
27. Ibid., 151.
28. Http://www.armoury.truvelo.co.za, (7 September 2003).
29. Heitman, *Unpublished Manuscript*, 151.
30. Ibid.

Bibliography

BOOKS AND ARTICLES

Abrahams, Diane. *Defence conversion in South Africa: A faded ideal?* Pretoria: Institute for Security Studies, ISS Occasional Paper 51, 2001.

Adam, H., and K. Moodley. *South Africa's Negotiated Revolution*. Cape Town: David Philip, 1992.

Adams, John. *The Unnatural Alliance*. London: Quartet Books, 1984.

Albright, David. "South Africa and the Affordable Bomb," *The Bulletin of Atomic Scientists* (July/August 1994): 37-47.

Alden, Chris. *Apartheid's Last Stand: The Rise and Fall of the South African Security State*. New York: St. Martin's Press, 1996.

Barrell, Howard. *MK: The ANC's Armed Struggle*. London: Penguin, 1990.

Batchelor, Peter, and Susan Willett. *Disarmament and Defence: Industrial Adjustment in South Africa*. London: Oxford University Press, 1998.

Beit-Hallahmi, Benjamin. *The Israeli Connection: Who Israel Arms and Why*. New York: Phanton Books, 1987.

Bitzinger, Richard A. *Towards a Brave New Arms Industry*? New York: Oxford University Press, IISS Adelphi Paper 356, 2003.

Botha, David. *Offsetting the Costs of SA's Strategic Defence Packages*. Pretoria: Institute for Security Studies, ISS Occasional Paper 75, 2003.

– *South Africa's Defence Industry: Charting a New Course?* Pretoria: Institute for Security Studies, ISS Occasional Paper 78, 2003.

– "The Arms Deal Controversy," *African Security Review* 12, no. 3 (2003): 99-103.

Breytenbach, Jan. *The Buffalo Soldiers: The story of South Africa's 32-Battalion 1975-1993*. Alberton, South Africa: Galago, 2002.

Bridgland, Fred. *The War for Africa: Twelve Months that Transformed a Continent*. Gibraltar: Ashanti Publishing Limited, 1990.

Brummer, Stefaans. "SA Arms-Dealing Underworld." (Johannesburg) *Weekly Mail & Guardian*, 2-8 June 1995, 9.

Brzoska, M. "Arming South Africa in the shadow of the UN arms embargo" *Defense Analysis* 7, no. 1 (1999): 21-38.

Burgess, Stephen, and Helen Purkitt. *The Rollback of South Africa's Chemical and Biological Warfare Program*. Maxwell Air Force Base, AL: USAF Counterproliferation Center, Air War College, 2001.

Cawthra, Gavin. *Brutal Force: The Apartheid War Machine*. London: International Defence and Aid Fund for Southern Africa, 1986.

– "South Africa." Pp. 142-179, in *Arms Procurement Decision Making, Volume II: Chile, Greece, Malaysia, Poland, South Africa and Taiwan*, edited by Ravinder Pal Singh. Oxford: Oxford University Press, 2000.

Chalmers, Robyn and Hopewell Radebe. "Arms deal saga is sending wrong messages to the world." (Johannesburg) *Business Day*, 29 August 2003.

Cilliers, Jakkie. *Defence Acquisitions-Unpacking the Package Deals*. Pretoria: Institute for Security Studies, ISS Occasional Paper 29, 1998.

– *The South African Army in Transition*. Pretoria: Institute for Security Studies, ISS Occasional Paper Number 26, 1998.

Cilliers, Jakkie, and Hussein Solomon, Editors. *People, Poverty and Peace: Human Security in Southern Africa*. Pretoria: Institute for Security Studies, ISS Monograph Number 4, 1996.

Clough, Michael and Jeffrey Herbst. *South Africa's Changing Regional Strategy: Beyond Destabilization*. New York: Council on Foreign Relations, 1989.

Cobbett, William, Daryl Glaser, Doug Hindson, and Mark Swilling. "A Critical Analysis of the South African State's Reform Strategies in the 1980s." Pp. 19-51 in *State, Resistance and Change in South Africa*, edited by P. Frankel, N. Pines and M. Swilling. London: Croom Helm,1988.

Cock, Jacklyn, and Laurie Nathan, Editors. *War and Society: The Militarisation of South Africa*. Cape Town: David Philip, 1989.

Cohen, Stephen P. *The Idea of Pakistan*. Washington, DC: Brookings Institute Press, 2004.

Cohen, Tim. "Arms-deal Report Cover-up." (Johannesburg) *Business Day*, 7 January 2005.

Cohen, Tim and Nicola Jenvey. "Court gives go-ahead for R150m arms claim." (Johannesburg) *Business Day*, 31 January 2005.

Davies, Robert, and Dan O'Meara. "Total Strategy in Southern Africa – An Analysis of South African Regional Policy Since 1978." Pp. 179-217 in *Exporting Apartheid: Foreign Policies in Southern Africa 1978-1988*, edited by S. Chan. New York: St. Martin's Press, 1990.

Dawes, Nic. "No plane, no gain." (Johannesburg) *Mail and Guardian*, 9 December 2004.

De Klerk, F.W. *The Last Trek – A New Beginning*. New York: St. Martin's Press, 1999.

De Villiers, F.W., Roger Jardine and Mitchell Reiss. "Why South Africa Gave Up the Bomb." *Foreign Affairs* 72 (November-December 1993): 98-109.

El-Abd, Hesham, and Michael Kenney O'Sullivan. "Encountering Countertrade." *Journal of Defense and Diplomacy* 2, no. 6 (1984).

Ellert, H. *The Rhodesian Front War*. Gweru, Zimbabwe: Mambo Press, 1993.

Evans, Michael. "Deadly Rockets Blast Way Through." *London Times*, 10 November 2004.

Fanaroff, B., "A trade unionist perspective on the future of the armaments industry in South Africa." *South African Defence Review* 7 (1992):10-4.

Ferreira, Rialize. "The integration of the armed forces into the South African National Defence Force: a sociological investigation." Pp. 82-110 in *Armed Forces and the International Diversities*, edited by Leena Parmar. Jaipur, India: Pointer Publications, 2001.

Fig, David. "Apartheid's Nuclear Arsenal: Deviation from Development." Pp. 163-180 in *From Defence to Development: Redirecting Military Resources in South Africa*, edited by. J. Cock and P. Mckenzie. Cape Town: David Phillip, 1998.

Frankel, Phillip H. *Pretoria's Praetorians*. Cambridge: Cambridge University Press, 1984.

Friedman, Steven, Editor. *The Long Journey: South Africa's Quest for a Negotiated Settlement*. Johannesburg: Ravan Press, 1993.

Geldenhuys, Deon. *The Diplomacy of Isolation: South African Foreign Policy Making*. New York: St. Martin's Press, 1984.

– *Isolated States: A Comparative Analysis*. Johannesburg: Jonathan Ball Publishers, 1990.

Giliomee, Hermann. *The Afrikaners: Biography of a People (Reconsiderations in Southern African History)*. Richmond, VA: University Press of Virginia, 2003.

Gleijeses, Piero. *Conflicting Missions: Havana, Washington, Pretoria*. Alberton, South Africa: Galago Books, 2003.

Gould, Chandré and Peter I. Folb. "The South African Chemical and Biological Warfare Program: An Overview." *The Nonproliferation Review* 7, no. 3 (Fall-Winter 2000): 10-23.

Grundy, Kenneth W. *Soldiers without Politics: Blacks in the South African Armed Forces*. Berkely, CA: University of California Press, 1983.

– *The Militarisation of South African Politics*. London: I. B. Tauris, 1986.

Hamann, Hilton. *Days of the Generals*. Cape Town: Zebra Press, 2001.

Hammond, Grant T. "Offset, Arms and Innovation." *The Washington Quarterly* (Winter 1987): 173-185.

– *Countertrade, Offsets and Barter: An International Political Economy*. London: Pinter Publishers Ltd, 1990.

Hanlon, J. *Apartheid's Second Front: South Africa's War Against its Neighbours*. London: Penguin, 1986.

Harper, Paddy. "Shaik Letters Drop Zuma's Name." (Johannesburg) *Sunday Times*, 7 February 2005.

Hartley, Wyndham. "Arms package faces critical risk as funds dry up." (Johannesburg) *Business Day*, 2 February 2005.

Harvey, Robert. *The Fall of Apartheid: The Inside Story from Smuts to Mbeki*. New York: Palgrave, 2001.

Heitman, Helmoed-Römer. *South African War Machine*. Novato, CA: Presidio Press, 1985.

– *War in Angola*. Gibraltar: Ashanti Publishing, Limited, 1990.

– "Opportunity Knocks." *Janes Defence Weekly*, 3 August 2005, 24-9.

Herlihy, James. "South Africa's Arms Exporter Seeks International Respectability." *New York Times*, 26 August 1994, 9-13.

Hersh, Seymour M. *The Samson Option*. New York: Random House, 1991.

Hough, M., "The UN arms embargo against South Africa: an assessment," *Strategic Review for Southern Africa* 10, no. 2 (1988): 22-38.

Hounam, Peter and Steve McQuillen. *The Mini-Nuke Conspiracy: Mandela's Nuclear Nightmare*. London: Faber and Faber, 1995.

Horton III, Roy E. *Out of (South) Africa: Pretoria's Nuclear Weapons Experience*. Colorado Springs, CO: USAFA Institute for National Security Studies, Occasional Paper Number 27, 1999.

Howlett, D., and J. Simpson. "Nuclearisation and denuclearisation in South Africa," *Survival* 35, no. 3 (1993): 154-73.

Jaster, Robert S. "South African Defense Strategy and the Growing Influence of the Military." Pp. 121-52 in *Arms and the African*, edited by W. Foltz and H. Bienen. New Haven, CT: York University Press, 1985.

– "Pretoria's Nuclear Diplomacy," *CSIS Africa Notes* Number 81 (22 January 1988).

Joseph, Benjamin M. *Beseiged Bedfellows: Israel and the Land of Apartheid*. New York: Greenwood Press, 1998.

Kasrils, Ronnie. "The Future of South Africa's Defence Industry: the Government's Perspective." Pp. 117-126 in *South Africa's Defence and Security in the 21st Century*, edited by W. Gutteridge. Brookfield, VT: Dartmouth Publishing Company, 1996.

Kent, Vanessa and Mark Malan. *Decisions, Decisions: South Africa's foray into regional peace operations*. Pretoria: Institute for Security Studies, ISS Occasional Paper 72, 2003.

Kingwill, Denys Graham. *The CSIR-the first 40 Years*. Pretoria: Scientia Printers, CSIR, 1990.

Klare, M. T. *US Arms Deliveries to South Africa*. Washington: Transnational Institute, 1977.

Klinghoffer, Arthur Jay. *The Angolan War: A Study in Soviet Policy in the Third World*. Boulder, CO: Westview Press, 1980.

Lambert, Mark. "Whites must offer fair share." *Sunday Times Business Times*, 7 March 2004, 3.

Landgren, Signe. *Embargo Disimplemented: South Africa's Military Industry*. New York: Oxford University Press, 1989.

Laufer, Stephen and Arthur Gavshon. "The Real Reason for South Africa's Nukes." (Johannesburg) *Weekly Mail & Guardian*, 1 April-26 March 1993, 3.

Le Roux, Len, "The South African National Defence Force and its involvement in the Defence Review process." Pp. 153-169 in *Ourselves to Know*, edited by R. Williams, G. Cawthra and D. Abrahams. Pretoria: Instititute for Security Studies, 2003.

Leysens, A. "South Africa's military strategic link with Latin America: past developments and future prospects." *International Affairs Bulletin* 15, no. 3 (1991): 23-47.

Liberman, Peter. "The Rise and Fall of the South African Bomb." *International Security* 26, no. 2 (Fall 2001): 45-86.

Liebenberg, Ian, and Rialize Ferreira. "Loyal service and yet "demobbed" – Demobilization and the economic reintegration of South Africa's demobilized military personnel." *Journal of Asian and African Studies* 37, no. 3-5 (December 2002): 299-317.

Lowther, William. *Arms and the Man: Dr. Gerald Bull, Iraq and the Supergun*. Novato, CA: Presidio Press, 1991.

Mann, Michael. "The Giant Stirs: South African Business in the Age of Reform." Pp. 52-86 in *State, Resistance and Change in South Africa*, edited by in P. Frankel, N. Pines, and M. Swilling. London: Croom Helm, 1988.

Marcum, John. "Retrenchment and Recalculation: South Africa and the Anglo-Namibian Agreements." Pp. 131-48 in *Disengagment from Southwest Africa: The Prospects for Peace in Angola and Namibia*, edited by O. Kahn. New Brunswick, ME: Transaction Publishers, 1991.

Martin, Stephen, Editor. The *Economics of Offsets: Defence, Procurement and Countertrade (Studies in Defence Economics)*. Amsterdam: Routledge, 1996.

Matthews, R. "The Development of the South African military industrial complex." *Defence Analysis* 4, no. 1 (1988): 7-24.

McMillan, S. "Economic Growth and Military Spending in South Africa." *International Interactions* 18, no. 1 (1992): 35-50.

McWilliams, James P. *Armscor: South Africa's Arms Merchant*. London: Brasseys, 1989.

Mills, Greg. "The South African National Defence Force: Between Downsizing and New Capabilities." *Naval War College Review* LII, no. 1 (Winter 1999): 78-99.

Mills, G., and S. Baynham. "South African Foreign Policy, 1945-1990." Pp. 18-33 in *From Pariah to Participant*, edited by G. Mills. Johannesburg: South African Institute of International Affairs, 1994.

Minter, William. *Apartheid's Contras: An inquiry into the roots of war in Angola and Mozambique*. Johannesburg: Witwatersrand University Press, 1994.

Moll, T. "From booster to brake? Apartheid and economic growth in perspective." Pp. 73-87 in *The Political Economy of South Africa*, edited by N. Nattrass and E. Ardington. London: Oxford University Press, Press, 1990.

– " Did the apartheid economy fail?" *Journal of Southern African Studies* 17, no. 2 (1991): 271-91.

Moorcraft, Paul L. *Africa's Superpower*. Johannesburg: Sygma Books, 1981.

Mufson, Steven. *Fighting Years: Black Resistance and the Struggle for a New South Africa*. Boston: Beacon Press, 1990.

Muldoon, Timothy J. "South Africa: A Case Study in Nuclear Proliferation and Nuclear Rollback." *National Security Studies Quarterly* 5, no. 2 (Spring 1999): 83-106.

Nathan, Laurie. *The changing of the guard: armed forces and defence policy in a democratic South Africa*. Cape Town: HSRC Publishers, 1994.

– "Beyond arms and armed forces: a new approach to security." *South African Defence Review* 4 (1992): 12-21.

"National Report: Top ANC Official Arrested." *Financial Gazette*, 4 October 2001.

Newby-Fraser, A. R. *Chain Reaction*. Pretoria: Atomic Energy Board, 1979.

Ngoma, Naison. *Hawks, Doves or Penguins? A critical review of the SADC military intervention in the DRC*. Pretoria: Institute for Security Studies, ISS Paper 88, 2004.

"NUMSA Denies Arms Deal Payoff." *Cape Argus*, 1 December 1999.

"Obituary: Richard Charter." (Johannesburg) *Sunday Times*, 1 February 2004.

Ottaway, David. *Chained Together: Mandela, De Klerk and the Struggle to Remake South Africa*. New York: Random House, 1993.

Pabian, Frank V. "South Africa's Nuclear Weapons Program: Lessons for Nonproliferation Policy." *The Nonproliferation Review* 3, no. 1 (Fall 1995): 11.

Phasiwe, Khulu. "Denel Primes Itself to Offload Workers." (Johannesburg) *Business Day*, 10 February 2005.

Price, Robert M. "Pretoria's Southern African Strategy." Pp. 145-69 in *Exporting Apartheid: Foreign Policies in Southern Africa 1978-1988*, edited by S. Chan. New York: St. Martin's Press, 1990.

Purkitt, Helen E., and Stephen F. Burgess. *South Africa's Weapons of Mass Destruction*. Bloomington, IN: Indiana University Press, 2005.

Ratcliffe, S. "The shifting dominance: the foundations of the armaments industry during the 1960s." *Africa Perspective* 25 (1984): 24-39.

Rotberg, R. "Decision making and the military in South Africa." Pp. 13-26 in *South Africa and its Neighbors: Regional Security and Self Interest*, edited by R. Rotberg. Lexington, MA: Lexington Books, 1985.

"SADC Report Modise's Dynamic Drawn into SA Arms Scandal." (Cape Town) *Financial Times*, 6 April 2001.

"SA Politics: Ambassadors quizzed over arms procurement kick-back claims." (Cape Town) *Financial Times*, 15 November 1999.

Sagan, Scott D. "Why do States Build Nuclear Weapons? Three Models in Search of a Bomb." *International Security* 21, no. 3 (Winter 1996/97): 69-71.

Schmidt, Michael. "Proof of SA Nuclear Plan." (Johannesburg) *Sunday Times*, 12 October 2003: 5.

Seegers, A. "Apartheid's military: its origins and development." Pp. 143-172 in *The State of Apartheid*, edited by W. James. Boulder, CO: Lynne Rienner, 1987.

– "South Africa's national security management system, 1972-1990." *Journal of Modern African Studies* 29, no. 2 (1991): 253-73.

– *The Military in the Making of Modern South* Africa. London: I. B. Tauris, 1996.

Shaw, M. "Biting the bullet: negotiating democracy's defence." Pp. 228-56 in *South African Review 7: The Small Miracle. South Africa's Negotiated Settlement*, edited by S. Friedman and D. Atkinson. Johannesburg: Ravan Press, 1994.

Shaw, T., and E. Leppan. "South Africa: white power and the regional military-industrial complex," Pp. 251-77 in *Southern Africa in the 1980s*, edited by O. Aluko and T. Shaw. London: George Allen & Unwin, 1985.

Shearer, J. "Denuclearisation of Africa: the South African dimension." *Disarmament* 16, no. 2 (1993): 171-85.

Sole, D. "South African Foreign Policy Assumptions and Objectives from Herzog to De Klerk." *South African Journal of International Affairs* 2, no.1 (1994): 104-13.

Solomon, Hussein. "From marginalised to dominant discourse: Reflections on the evolution of new security thinking" in *Caring Security in Africa*, edited by H. Solomon and M. van Aardt. Pretoria: Institute for Security Studies, ISS Monograph No. 20, 1998.

Solomon, H. and M. van Aardt, Editors, *Caring Security in Africa*, Pretoria: Institute for Security Studies, ISS Monograph No. 20, 1998.

"South Africa Confirms Super LYNX 300 Order." *Military Technology* 27, no. 10 (October 2003): 98.

Sparks, Allister. *The Mind of South Africa: The Story of the Rise and Fall of Apartheid*. London: Mandarin, 1990.

Stander, Siegfried. *Like the Wind: The Story of the South African Army*. Cape Town: Saayman & Weber, 1985.

Steenkamp, Willem. *The Soldiers*. Cape Town: Don Nelson, 1978.

– *Borderstrike! South Africa into Angola*. Durban: Butterworths, 1983.

Steyn, Hannes, Richard van der Walt, and Jan van Loggerenburg. *Armament and Disarmament: South Africa's Nuclear Weapons Experience*. Pretoria: Network Publishers, 2003.

Stockwell, John. *In Search of Enemies: A CIA Story*. New York: W.W. Norton & Co., Inc, 1984.

Stumpf, Waldo. "South Africa's Nuclear Weapons Program: From Deterrence to Dismantlement." *Arms Control Today* 25, no. 10 (December 1995/January 1996): 4-7.

Thayer, Bradley A. "The Causes of Nuclear Nonproliferation and the Utility of the Nuclear Nonproliferation Regime." *Security Studies* 4, no. 3 (Spring 1995): 449-94.

Thompson, Scott. "South Africa and the 1988 Agreements." Pp. 117-30 in *Disengagment from Southwest Africa: The Prospects for Peace in Angola and Namibia*, edited by O. Kahn. New Brunswick, ME: Transaction Publishers.

Thornycroft, Petra. "SA's Poison Gas Secrets Sold to Libya," (Johannesburg) *Mail & Guardian*, 7 January 1997.

Van Aardt, M. "Doing battle with security: a Southern African approach." *South African Journal of International Affairs* 3, no. 2 (Summer 1996): 13-28.

Van der Waals, Willem. *Portugal's War in Angola, 1961-1974*. Rivonia, South Africa: Ashanti Publishing, 1993.

Vatcher, William H. *White Laager: The Rise of Afrikaner Nationalism*. New York: Praeger, 1965.

Williams, Rocky. "Defence in a democracy: the South African Defence Review and the redefinition of the parameters of the national defence debate." Pp. 205-23 in *Ourselves to Know*, edited by R. Williams, Gavin Cawthra and D. Abrahams. Pretoria: Institute for Security Studies, 2003.

DOCUMENTS, PAPERS AND ON-LINE SOURCES

"26 January 2004: Newspaper continues anti-Hawk campaign despite official displeasure." Http:www.saairforce.co.za/news.htm, (15 January 2005).

"$5.2bn military deal defended against backdrop of regional war," Financial Times information, 1 September 1998, Southscan. Http://www.southscan.gn.apc.org, (23 August 2003).

Armaments Corporation of South Africa, Limited Act, 2003. Http://www.info.gov.za/documents/acts/index.htm#2003, (21 March 2005).

ARMSCOR *Annual Report*, 2002-2003. Http://www.armscor.co.za, (12 September 2004).

– *Annual Report*, 2003-2004. Http://www.armscor.co.za, (26 March 2005.

"Arms deal challenged in Court." South African Press Association News Release, 20 March 2003. Http://iafrica.com.news/sa/220753.htm, (22 September 2004).

Batchelor, Peter, and Paul Dunne. "Industrial Participation, Investment and Growth: The Case of South Africa's Defence Related Industry." Paper prepared for the Centre for Conflict Resolution, Cape Town and Middlesex University Business School, United Kingdom, February 2000.

Braun, Gedahlia. "South Africa Under Black Rule (Part 1)." *American Renaissance* 9, no. 7 (July 1988). Http://www.amren.com.987issue/987issue. html, (10 January 2005).

Christie, Renfrew. "South Africa's Nuclear History." Paper presented at the Nuclear History Program Fourth Annual Conference, Nice, France, 23-27 June 1993.

Constitution of the Republic of South Africa, 1996. Act No. 108 of 1996. Http://www.info.gov.za/documents/acts/index.htm#1996, (25 March 2005).

Denel Annual Report 2002. Http://www.denel.co.za, (25 November 2003).
– *Annual Report 2003*. Http://www.denel.co.za, (6 February 2004).
– *Annual Report 2004*. Http://www.denel.co.za, (31 March 2005).

Englebrecht, Leon. "Developments in the South African Defence Industry: International Alliances." Summary of NIC/ISS Conference: Transformation in Global Defence Markets and Industries, 4-5 November 2000. Http://sss.arms deal-vpo.co.za/special_items/reading/developments_sa_defence_html., (23 April 2004).

Executive Summary, Republic of South Africa-United States of America: 7th Annual Defense Committee Meeting: 4-6 June 2003, Annapolis, Maryland. Http://www.defenselink.mil/policy/isa/africa/DEFCOM03minutes.htm, (21 March 2005).

Heitman, Helmoed-Römer. *Unpublished Manuscript*. Provided to the author in June 2003.

Human Rights Watch Report, 12, no. 5 (A) (October 2000). Http://www.hrw.org/reports/2000/safrica/Sarfio00-05.htm), (25 March 2005).

"Lekota signs order for four naval helicopters." South African Press Association Global News Wire, 15 August 2003. Http:www.sapa.org.za, (15 August 2003).

"Mechem," *Journal of Mine Action*, June 2000. Http://www.maic.jmu.edu/journal/ 4.2/Profiles/mechem.htm, (10 July 2004).

Mills, Greg. *South Africa: The Total National Strategy and Region Policy During the Botha Years, 1978-1989*. Doctoral Dissertation, University of Lancaster, UK, 1990.

Niesen, Cpl. Joe. "New vehicles helping against IEDs." ARNEWS, 11 February 2004. Http://www4.army.mil, (28 October 2004).

Nyanda, S. "The South African National Defence Force and Peace Missions in Africa." Paper presented at the African Dialogue Series, University of Pretoria, 27 February 2003.

Pather, Sashi. "How SA-made vehicle saved US troops." (Johannesburg) *Sunday Times*, 24 October 2004. Http://www.sundaytimes.co.za, (28 October 2004).

"Ready to Govern: ANC Policy Guideline for a Democratic South Africa," May 1992. Http://www.anc.org.za/ancdocs/history/readyto.html, (25 March 2005).

Report by the Director General: The Agency's Verification Activities in South Africa. Vienna: International Atomic Energy Agency, 1993.

Reunert Annual Report – 2003. Http://www.reunert.com/invester/annual_2003/, (25 April 2004)

"SA firm in R468m deal to supply vehicles to US army." Creamer Media, 2005/02/25. Http://www.miningweekly.co.za, (15 March 2005).

"SAFRICA: Defence Minister 'Quietly' Signs Deal of Supply of Naval Helicopters." Global News Wire-Asia Africa Intelligence Wire, BBC Monitoring International Reports, 15 August 2003.

"Shaik in Court, Released on $1,000 Bail." South African Press Association News Release, 25 August 2003. Http://www.sapa.za.org, (25 August 2003).

Sokolski, Henry. "Ending South Africa's Rocket Program: A Nonproliferation Success." Nonproliferation Policy Education Center, Washington DC, Fall 1993. Http://www.npec-web.org/essay/southafrica.htm, (2 May 2005).

Sole, Donald B. "The South African Nuclear Case in the Light of Recent Revelations." Paper presented at the conference, New Horizons in Arms Control and Verification, Southern Methodist University, Dallas, Texas, October 1993.

"South Africa: An Irresponsible Arms Trader?" *Global Dialogue* 4 (2 August 1999). Http://www.igd.org.za/pub/g-dialogue/Special_feature/ trader.html, (3 May 2005).

"South African Defence Industry." Global Security Organization. Http://www.globalsecurity.org/military/world/rsa/industry.htm, (30 June 2003).

"South Africa gets ready to announce fighter trainer winner." Reed Business Information Ltd., 14 October 1998. Http://www.reedbusiness.com, (23 August 2003).

South African National Defence Force Military Strategy. Pretoria: Department of Defence, Printed by 1 Military Training Regiment, 2002.

South African Defence Review (1996). Pretoria: Department of Defence, Printed by 1 Military Printing Regiment, 1998.

South African White Paper on Defence (1998). Pretoria: Department of Defence, Printed by 1 Military Printing Regiment, 1998.

South African White Paper on the South African Defence-Related Industries (1999). Http://www.gov.za/whitepaper/1999/defence.htm, (22 February 2005).

South Africa's Nuclear Weapons Program: An Annotated Chronology, 1969-1994. Http://www.cns.miis.edu/research/safrica/chron.htm, (3 May 2005).

Stoppard, Anthony. "Politics-South Africa: ANC Faces Corruption in its Ranks." Global Information Network, 30 August 2003. Http://proquest.umi.com, (26 March 2005).

United Nations Development Program. "Human Development Report, 1994." Http://hdr.undp.org/reprts/global/1994/en, (12 March 2004).

U.S. Arms Transfers to South Africa in Violation of the United Nations Voluntary Arms Embargo, 1963-1977. New York: UN Centre Against Apartheid, 1978.

U.S. *Federal Register* 63, no. 42 (4 March 1998), notices, Public Notice 2752, p. 10671.

U.S. *Federal Register* 69, no. 139 (21 July 2004), Notices, Public Notice 4768, p. 43656.

OFFICIAL WEB SITES

Advanced Technologies and Engineering (ATE). Http://www.ate-aerospace.com, (15 May 2004).
Aerospace Monitoring Systems (AMS). Http://www.ams.co.za, (20 July 2004).
Aerosud. Http:www.aerosud.co.za, (20 May 2004).
African Defence Systems (ADS). Http://www.ads.co.za, (27 April 2004).
African National Congress. Http://www.anc.org, (25 March 2005).
AMD. Http://www.amd.org.za, (6 May 2005).
AMS. Http://www.ams.co.za, (12 May 2004).
ANSYS Integrated Systems. Http://www.ansys.co.za , (12 July 2004).
Ariel Technologies. Http://www.ariel.co.za (6 September 2003).
Armscor. Http://www.armscor.co.za, (6 May 2005).
BAE Land Systems OMC. Http://www.baelandsystemsomc.ca.za, (6 May 2005).
C^2I^2. Http://www.ccii.co.za, (15 May 2004).
CSIR. Http://www.csir.co.za, (6 May 2005).
Cybersim. Http://www.cybersim.co.za (6 September 2003).
Denel. Http://www.denel.co.za, (6 May 2005).
Diel. Http://www.diel.co.za (7 September 2003).
Electromagnetic Software and Systems (EMS). Http://www.emss.co.za (7 September 2003).
Fuchs. Http://www.fuchs.co.za, (13 March 2004).
General Dynamics Land Systems (GDLS). Http://www.gdls.com, (12 September 2004).
Gripen (Saab multirole fighter aircraft part of the South African Defence Packages). Http://www.gripen.com, (6 May 2005).
Grintek. Http://www.grintek.com, (15 May 2004).
Human Rights Watch. Http://www.hrw.org, (25 March 2005).
MGL Milkor Marketing. Http://www.milkor.co.za (12 September 2003).
Nanoteq. Http://www.nonteq.com, (7 September 2003).
Offsets Information Website, University of Plymouth. UK, Http://www.plymouth.ac.uk/politics, (3 March 2005).
Reunert. Http://www.reunert.co.za, (22 May 2004).
Safair. Http://www.safair.co.za, (20 July 2004).
South African Arms Exhibitions. Http://www.aadexpo.co.za, (15 March 2005).
South African Government. Http://www.gov.za, (25 March 2005).
South African Government White Papers. Http://www.info.gov.za/whitepapers/index.htm, (25 March 2005).
Stingray Marine. Http://www.stingramarine.com (6 September 2003).
Tellumat. Http://www.tellumat.com, (11 September 2003).
Truvelo Armoury. Http://www.armoury.truvelo.co.za (7 September 2003).
United Nations Development Program/Human Development Report. Http://www.hdr.undp.org, (25 March 2005).

U.S. Bureau of Alcohol, Tobacco and Firearms (BATF). Http://www.atf.gov, (6 May 2005).

U.S. Defense Security Cooperation Agency (DSCA). Http://www.dsca.mil, (22 March 2005).

U.S. Foreign Military Assistance Funding. Http://www.fas.org, (25 March 2005).

INTERVIEWS

Cawthra, Gavin. Director, Centre for Defence and Security Management, Graduate School of Public & Development Management, University of the Witwatersrand, Johannesburg, 19 June 2003 and 22 June 2004.

Cilliers, Jakkie. Executive Director, Institute for Security Studies, Pretoria, 11 March 2004 and 18 June 2004.

Ferreira, Realize M. Professor of Sociology, University of South Africa, Pretoria. 11 July 2005.

Forney, Rex (Colonel, Military Police Corps, US Army). Montgomery, Alabama, 18 and 29 April 2005.

Gerber, Paul. Grintek Group Manager for Business Coordination and Counter Trade, Pretoria, 13 June 2003 and 21 June 2004.

Heitman, Helmoed-Römer. South African Correspondent for Jane's Defence Weekly, Cape Town, 6 June 2003 and 23 June 2004.

Holloway, Alan. Vice Chairman of AMD and Managing Director ANSYS, Pretoria, 21 June 2004.

Lang, Duncan. Contractor to the US Secretary of Defense, Washington, DC, 12 March 2005.

Le Roux. Len (Major General, South African Air Force, Retired). Head of the Defence Sector Programme, Institute for Security Studies, Pretoria, 11 March 2004 and 18 June 2004.

Miner, James (Lieutenant Colonel, US Air Force). US Defense Security Cooperation Agency, Washington, DC. 16 March 2005.

Nepgen, Andre. Director of CSIR/Defencetek, Pretoria, 21 June 2003.

Ngwenya, Reginald. Director of AMD, 21 June 2004.

Oates, Arthur J. R. Director of Business Development, Denel, Pretoria, 12 June 2003.

Roux, Andre. Director of the Futures Institute, Cape Town, 23 June 2003.

Rupiya, Martin. (Lieutenant Colonel, Zimbabwe National Army, retired). Senior Researcher, Defence Sector Programme, Institute for Security Studies, Pretoria. 11 March 2004.

Savides, Tony (Brigadier, South African Army, Retired). Director of International Marketing, BAE Land Systems OMC, Benoni, South Africa, 18 July 2003.

Steyn, Hannes (former Senior General Manager, Research and Development, Armscor). Pretoria, 11 June 2005.

Thackwray, Winston (Major General, South African Air Force, Retired). Advisor to the Managing Director EADS South Africa, Pretoria, 21 June 2004.

Turton, Anthony R. Council for Scientific and Industrial Research, Pretoria, 22 June 2004.

Van Rensburg, Johan. Commercial Specialist, US Commercial Service, Johannesburg, 23 June 2004.

Note: Between June 2003 and July 2005, the author interviewed approximately seventy serving members of the South African National Defence Force and approximately forty additional mid-level arms industry managerial or research personnel who requested that their names not be cited in this study.

Index

Z

About the Author

Dan Henk currently serves as a social anthropologist at the US Air War College, where his work overlaps the topics of culture and national security. His recent research explores the expanding worldwide definitions of "security" and the evolution of new military roles and missions.

He attained his doctorate in Anthropology from the University of Florida in 1988 and is a graduate both of the US Army Command and General Staff College and US Army War College.

From 1970 to 1999 Henk served in the Regular Army of the United States, retiring in the rank of Colonel. His military assignments included combat in Viet Nam and Grenada, service in the United States, Europe, Asia and Africa, and accreditation as military attaché to four African countries. During his military career, he taught at the US Military Academy, John F. Kennedy Special Warfare Center, Joint Military Intelligence College, and US Army War College. When he retired from the Army in 1999, he was that service's senior specialist on African affairs. He has published and lectured widely on issues of African security and US regional policy.

www.ingramcontent.com/pod-product-compliance
Ingram Content Group UK Ltd.
Pitfield, Milton Keynes, MK11 3LW, UK
UKHW012039150926
475560UK00001B/85